Science Communication in a Crisis

Science Communication in a Crisis: An Insider's Guide identifies the principal challenges that scientists face when communicating with different stakeholder groups and offers advice on how to navigate the maze of competing interests and deliver actionable science when the clock is ticking.

If a scientist's goal is to deliver content and expertise to the people who need it, then other stakeholder groups—the media, the government, industry—need to be considered as partners to collaborate with in order to solve problems. Written by established scientist Christopher Reddy, who has been on the front lines of several environmental crisis events, the book highlights ten specific challenges and reflects on mistakes made and lessons learned. Reddy's aim is not to teach scientists how to ace an interview or craft a soundbite, rather, through exploring several high-profile case studies, including the *North Cape* oil spill, *Deepwater Horizon*, and the 2021 Sri Lanka shipping disaster, he presents a clear pathway to effective and collaborative communication.

This book will be a great resource for junior and established scientists who want to make an impact, as well as students in courses such as environmental and science communication.

Christopher Reddy is a leader in the study of marine pollution and the development of environmentally friendly industrial chemicals. A senior scientist in the Department of Marine Chemistry and Geochemistry at Woods Hole Oceanographic Institution and faculty member of the MIT/WHOI joint program in oceanography, Reddy has led numerous field operations along coastlines, in the open ocean, and at the bottom of the sea to conduct transformative research that crosses disciplines and guides policy decisions worldwide.

"As a boots-on-the-ground scientist with an active research lab, Reddy's voice is one that many of our colleagues will identify with. This insider's perspective fills a gap in the lineup of existing science communication books."

Andrew Dessler, *Director, Texas Center for Climate Studies;*
Reta A. Haynes Chair in Geosciences, Texas A&M

"Communicating science in a crisis is risky and challenging but incredibly important. This book is generously packed with decades of experience and wisdom from the front lines that can make us all better communicators in times of need. A must for every scientist's bookshelf."

Dr. Asha de Vos, *Founder/Executive Director, Oceanswell, Sri Lanka*

"Focusing on how science communication transpires in real time, Reddy's insight from the trenches of recent environmental crises provides scientists with a blueprint for success with a range of stakeholder groups, from responders to the affected public."

Barbara Sherwood Lollar, *Dr. Norman Keevil Chair in*
Ore Deposits Geology, University of Toronto

Science Communication in a Crisis

An Insider's Guide

Christopher Reddy

Routledge
Taylor & Francis Group

LONDON AND NEW YORK

earthscan
from Routledge

Cover image: © Natalie Renier and Riley Orlando

First published 2023
by Routledge
4 Park Square, Milton Park, Abingdon, Oxon OX14 4RN

and by Routledge
605 Third Avenue, New York, NY 10158

Routledge is an imprint of the Taylor & Francis Group, an informa business

© 2023 Christopher Reddy

British Library Cataloguing-in-Publication Data
A catalogue record for this book is available from the British Library

ISBN: 978-1-032-37779-7 (hbk)
ISBN: 978-1-032-37780-3 (pbk)
ISBN: 978-1-003-34187-1 (ebk)

DOI: 10.4324/9781003341871

Typeset in Times New Roman
by Taylor & Francis Books

To all the researchers, health-care professionals, and first responders who carried us through COVID-19 with their tireless service and perseverance.

Contents

Acknowledgments

Writing a book that spans some thirty years of personal interactions leads to a long list of acknowledgments. I have been lucky and privileged to have had the time, resources, and support to have been able to learn from, meet with, and befriend a wide range of people. Few have been as fortunate.

My editor, Christopher Pitts, patiently listened to me during numerous Zoom meetings, read through everything I wrote, asked the right questions, and helped me to create a book that otherwise would have remained no more than random scraps of papers on my desk, sitting beneath half-empty cans of Diet Coke.

On a personal level, I am indebted to my parents: Joe and Sandra Reddy. They were committed to their children and sacrificed so much to ensure that my two sisters, four brothers, and I all had better lives than they did growing up. My parents were polar opposites, self-made, and managed successful careers all while raising us—along with the menagerie of lost souls who crashed on the floor and joined us for dinner at the picnic table in our kitchen. My dad is a no-nonsense gentleman, tenacious, and forever striving for excellence. He constantly reminded us to do the right thing and contribute to society and along the way taught a master class on leadership. My mom is easily one of the smartest and most creative people I have ever known. While working as an ER nurse, she had a legendary ability to remain calm under pressure and always had great stories to tell after working double shifts over the weekend. While my siblings (and their spouses) may not know exactly what I do, they have always been amazingly supportive and loving: Maureen and Jon, Colleen and Ken, Joe and Cindy, Tom, Dennis, and Jeanne, and Ryan and Melissa, and the late Kathy Reddy.

I am not a big fan of stereotypical depictions of scientists on TV shows and in the cinema, but I do wish I had a DeLorean and the science and engineering talent of Doc Brown. My first trip would have been back to 1976 to see my ailing grandfather Conrad Bennett and spoil him with all of the newest electronic gadgets. He was a techie before it was cool and kickstarted my lifelong interest in understanding how things work. To my grandmother Marion's endless frustration, my grandfather taught me how to light a blow torch, solder a pipe joint, and complete an electrical circuit before I was seven years old.

I married up and benefitted from the warmth and kindness showered upon me by my in-laws, Anthony and Joan DeMello, and also gained my sister-in-law Sidney and her husband Sean, along with my brother-in-law Jared.

In interviews, you sometimes hear a person asked about the one teacher in their life whom they wish to thank. In my case, I had multiple teachers who got me to where I am. All of my formal education was in Rhode Island's public education system. I can recite lessons and experiments from the many science teachers in my life: Ms. Millette, Mr. Razza, Mr. Cruciani, Mr. Shallcross, Ms. Babcock, and Mr. Roske (Cranston Public Schools, K-12); Professors Marzzacco, Williams, James and Elaine Magyar, Huizenga, Cooley, Greene, Glanz, and Gilbert (Rhode Island College; BS).

I hit the jackpot when Professor James Quinn, my PhD advisor at the Graduate School of Oceanography, University of Rhode Island, took the time to meet me on my lunch break while I was working in industry. He shepherded me from a nonmatriculating student to defending a PhD four years and four months later. His selfless devotion to his students set a bar for mentoring and advising that I will never reach. On top of that, he is the best chemist I know. When I arrived at WHOI as a bright-eyed postdoc, three kings provided me with wisdom and encouragement: Tim Eglinton, John Hayes, and John Farrington.

I am proof positive to the old adage that you should surround yourself with people who are smarter than you. My colleagues and I have made great science together, doing our part to contribute to society. Through thick and thin, Collin Ward and Dave Valentine have held my hand, watched my back, and soothed my nerves. I have also been fortunate to work with my former graduate students Ana Lima, Helen White, Desiree Plata, Kristin Pangallo, Karin Lemkau, and Anna Walsh, and postdocs Greg Slater, Emma Teuten, Christoph Aeppli, Ananya Sen Gupta, New Bob Swarthout, Ulrich Hanke, Mike Mazzotta, Taylor Nelson, and Yanchen Sun.

I am grateful to Rich Camilli, Dana Yoerger, Ryan Rodgers, Mark Hahn, Asha de Vos, Rick Gaines, Glenn Frysinger, Greg Hall, Ed Overton, David Yoskowitz, Tracy Ippolito, Greg O'Neil, Catherine Carmichael, Jennifer Culbertson, Emily Peacock, Kelsey Gosselin, Mike Timko, Burch Fisher, Jagos Radovic, Kara Lavender Law, Phil Gschwend, John Stegeman, Judy McDowell, Matt and Christine Charette, Michael Moore, Will Pecue, Wade Bryant, Hedy Edmonds, and Lihini Aluwihare. Sean Sylva has not only been a friend for almost thirty years, but he also saved my life on a research cruise. I recall pivotal conversations with France Córdova, Naomi Oreskes, Juliette Kayyem, Juliet Eilperin, Kristin Ludwig, Liesel Ritchie, Arnold Howitt, Andy Hoffman, and Chris Flowers. Extra special callouts to postdoc Bryan James, who lent an invaluable hand assisting with this book; my science partner, Bob Nelson, whose magic in the laboratory and field has been the cornerstone of our work; and John Kessler, for his encouragement in the early stages of this project and tremendous insight when reviewing earlier drafts of this book.

I cannot think of many other places to work besides WHOI that would have provided me with the support I have received over the years. Thanks to former directors Bob Gagosian, Susan Avery, and Mark Abbott, and current director Peter de Menocal. Additional gratitude to Jim Luyten, Larry Madin, Rick Murray, Chris Land, Rob Munier, Peter Hill, Terry Schaaf, Dick Pittenger, Jim Yoder, Ken Buesseler, Jeff Seewald, Bernard Peucker Ehrenbrink, Dennis McGillicuddy, Don Anderson, David Scully, Stephen Hoch, Rod Berens, John Richardson, Susan Humphris, Frank and Lisina Hoch, Bob and Anne James, Jim and Ruth Clark, Bill Kealy, Bill Rugh, Andy Solow, Bill Curry, Court Clayton, Sam Harp, Kathy Patterson, Joanne Tromp, Suzanne Pelisson, Danielle Fino, Linda Cannata, Mary Murphy, Jenn Carter, Donna Mortimer, and the indispensable Mary Zawoysky.

In the aftermath of *Deepwater Horizon*, my mental and physical health deteriorated. Many thanks to Dr. James Alban, Dr. Julie Callanan, Holly Oberacker, and Tracey Goodwin for getting me back on my feet.

Many thanks to all of those folks who provided advice, input, and fact-checking for this book: Vern Asper, Arne Diercks, John Ryan, Mandy Joye, Debbie French-McCay, Jacqui Michel, Scott Stout, Terry Hazen, Mark Schrope, Jay Cullen, Ed Boyle, Russ Flegal, Bill Arnold, Paige Novak, Matt Simcik, Dave De Vault, Joann Burkholder, Bob Haddad, Rob Ricker, Chuck Hopkinson, John Ryan, Bruce Lewenstein, Tom Ryerson, Tim Meehan, Ellen Franconi, Michael Mann and Stephanie Murphy. Thanks to Barbara Sherwood Lollar for her endorsement and Andy Dessler not only for his endorsement but also for providing guidance on the publishing process.

My research has been funded by the National Science Foundation (NSF), Office of Naval Research, Environmental Protection Agency, Sea Grant, Department of Energy, GOMRI, Seaver Institute, LECO, and Schlumberger Doll Research. My former program manager at the NSF, Don Rice, recommended me as the Academic Liaison at the *Deepwater Horizon* Unified Command when tensions between responders and academia were at a breaking point.

When it comes to science communication, I am indebted to Jim Kent, Ken Kostel, and Tom Hayden. They assured me that hiring an editor was not cheating and have edited countless opinion pieces for me. These fine graphic artists took my crazy ideas and sketches and turned them into artwork: Katherine Spencer Joyce, Eric Taylor, Jack Cooke, Amy Caracappa, Riley Orlando, and the very talented Natalie Renier, whose illustrations are featured in this book. The talented author and illustrator Molly Bang gave me the courage to move past my imposter syndrome.

Venturing outside of the Ivory Tower is daunting, but I had it pretty easy with the assistance I received. Jane Lubchenco and the rest of the talented staff with the former Aldo Leopold Leadership Program (now part of Earth Leadership Program) provided me with training. While walking our dogs at the Crane Wildlife Area in 2008, John Holdren encouraged me to submit an article I was working on to *Science* (tip of the hat to Sherman and Stella), and Kate Madin suggested I call it "Scientist Citizens." Thad Allen has always been generous with his advice. Steve Lehmann gave me some of the best counsel on how the real world works and kept me from turning into a guerrilla geochemist. The late Peter Lord set a standard for decency and professional journalists. I appreciate the patience and interest I have received from countless reporters.

Beyond some of the names mentioned in this book, I recognize my science heroes who paid the price for doing the right thing: Antoine Lavosier, J. Robert Oppenheimer, and Rosalind Franklin. While not a scientist, I learned a lot from the late Anthony Bourdain whose culinary skills were far outpaced by his decency and openness to different people and cultures. I encourage scientists to watch Bourdain's show, *Parts Unknown*, and experience his genuine interest in people and the foods they make and eat. I only wish my book was as effective and inspiring as his series.

I'd like to thank Lonny Lippsett, whose writing and editing skills, passion for communicating science, and friendship have inspired me along my own journey. He is a phenomenal teacher, mentor, and friend.

Writing a book leads to countless unexpected realizations about history, who and why you are, how you have changed, and the defining moments in your life. Perhaps the greatest joy in this adventure has been seeing the impact of my wife, Bryce. I am so thankful for the willingness, patience, and understanding that she radiates every day of our life together. Her caring, thoughtful, and loving approach to all walks of people shines through on so many pages of this book. I am far from the cocky and brash elitist that she met in 2005 and a considerably better person because of her. My children are an endless source of entertainment, love, and inspiration: my creative William, tenacious Elliot, and inquisitive Flynn.

Introduction

What exactly is a crisis? According to Merriam-Webster, it is (1) the turning point for better or worse in an acute disease or fever; or (2) a difficult or dangerous situation that needs serious attention.[1]

For the purpose of this book, I define a crisis as an environmental event. A crisis does not have to be mega-disaster like a global pandemic or a Category 5 hurricane. Crises come in different shapes and sizes, and they impact our lives to varying degrees. Some are one-time events (earthquakes), others are existential in nature and imperceptible to the human eye (ocean acidification), while still others are small or local enough to feel like nothing more than a minor disturbance in the warp and weft of everyday life.

For a scientist, a crisis is something else as well. It is a moment of opportunity: a time when you can provide clarity and information, thereby helping to affect the outcome of the event in a positive manner.

To give an example: on the morning of July 2, 2021, a natural gas pipeline in the Gulf of Mexico ruptured and began leaking gas, which made its way to the surface of the ocean. A nearby electrical storm then ignited the gas, which resulted in a rather sensational phenomenon. It appeared as if the ocean had caught fire: a swirling, glowing eye of Sauron peering back at us from a seething pit of magma, bubbling up from the depths of the sea.

Social media users pounced on the image, which was compounded by the apparent absurdity of nearby first-response boats that were trying to douse the flames with ... seawater? Was this really happening? Understandably, the image went viral. Heat waves were wracking the Pacific Northwest, wildfires were racing across California—was this yet another example that the end was nigh?

As we now know, this was simply a ruptured gas pipeline, not a leaking well. Pipelines can be shut down with valves. Thus, it was not a disaster, but it was definitely an event. And that, in my mind, made it a crisis. In the immediate aftermath, many people let their imaginations run wild. Millions of gallons of oil had already gushed into the ocean during the 2010 *Deepwater Horizon* spill in the Gulf of Mexico. Perhaps this was a repeat.

I bring this up because this is the perfect example of how a scientist can add clarity to an uncertain situation. Despite the fact that many people on

DOI: 10.4324/9781003341871-1

social media continued to call it an oil spill, I knew this was not the case. I could see from the flames that it was burning much too cleanly to be oil. And what's more, I knew the location of the pipeline. This was an active drilling area, which meant the industry had the infrastructure and supplies to deal with the problem. In my mind—and I say this as a chemist who has devoted his entire career to studying ocean pollutants—there would be no long-term environmental consequences.

And so I tweeted: "#GulfOfMexicoFire unlikely to hurt ocean life. Nat gas not oil. Like buying a house, it's all about location. @Pemex very lucky to have people/technology nearby; it was a damaged pipeline & not a well deep below the ocean. Demands follow-up. @WHOI @nsf @NatGeo."

I was simply reminding everyone: "Hey, this is not as bad as it looks. In this particular instance, everything will likely be okay." Not long after my tweet, a reporter from *Wired* contacted me for an interview. Since it seemed like a good opportunity to add more content to a hot topic, I agreed to speak with her. In a matter of hours, the story went live, and my message reached a wider audience.

This is but one example of how a scientist can become involved in a crisis. While something like the *Deepwater Horizon* spill may be a once-in-a-lifetime disaster, these types of smaller environmental events happen all the time. Becoming involved and making a difference can be a relatively low-risk activity for a scientist. It can be as simple as posting a single tweet, and it might take less than an hour of your time. But you still need to understand the rules of the game. You still need to know how to get that message across.

This book is not prescriptive. My aim is not to teach scientists how to ace an interview or craft a soundbite. Those are skills best taught by journalists or communications professionals. Rather, it is a collection of anecdotes: a reflection on the mistakes I have made and the lessons I have learned—and continue to learn—over the course of my own career. I'll discuss successes too, like the "burning ocean" story, because, after all, successful communication is our ultimate goal.

When it comes to communication, many scientists have three default modes: the first is to not communicate at all. When a reporter calls, you simply ignore them. The second is that of the stilted lover. In this case, you are approached by a journalist after the publication of a paper, or for an opinion on a hot topic or policy issue. During the interview, you go into great detail and explain all of your research. Then the next day, when the article comes out, you cringe in horror at the headline and lede, come to the conclusion that the uninformed journalist never understood you to begin with, and vow to never speak to the media again. The third is the role of sheriff. In this case you make use of your knowledge to challenge the actions or statements of industry or government. Do we need sheriffs? Absolutely. But does every interaction with industry and government have to be confrontational? No.

In this book, I am going to advocate for another communication mode: teamwork. If your end goal is delivering content and expertise to the people who need it, you will have to be part of a team—and this team will include people who are not scientists. How will you communicate with firefighters, the police force, the Coast Guard? With local government officials? With journalists who are not on their regular beat? As we will see, you cannot serve the same one-pot meal to every single stakeholder group. You will need to tailor your communication style depending on whom you are talking to.

This lesson has been made painfully clear to me on multiple occasions, but never more so than during the *Deepwater Horizon* oil spill. Although this event was unprecedented in its magnitude, it crystallizes so many of the challenges that scientists face, and is such a compelling story, that it has gone on to play a major role in this book.

And while you may not be an oceanographer or a chemist, I believe that the communication challenges I have faced over the course of my career will be the same ones you will face over the course of yours. The lessons I learned are also applicable to people on the other side of the fence: responders, journalists, engineers, industry contractors, activists, and government officials. Even if you might be thinking to yourself, *I'll never be involved in a major crisis like Deepwater Horizon*, no matter what you do, at some point in your life you will be asked to provide expertise. And in doing so, you'll likely find yourself facing these same challenges.

So what are the challenges that are specific to science communication? I have identified ten principal themes.

1 Career
2 Culture
3 Network
4 Speaking Out
5 Competition
6 Process
7 Misinformation
8 Legal
9 Impact
10 Teamwork

Each chapter in this book focuses on one of these particular communication challenges. As scientists, the primary challenge we have to navigate is our *career.* How many successful PhDs get a tenure-track position in the United States? According to *Science*, in 2017 that number was only 23%.[2] Traditionally, academic success has hinged more on getting published and securing funding for your research, and less on public outreach. Peer approval is also important. If you act in ways that are traditionally frowned upon in the scientific community—such as attracting too much media attention—the consequences to your career can be real.

Culture boils down to an awareness that each stakeholder group—responders, industry, policymakers, the media, the public, non-governmental organizations (NGOs)—is different. These stakeholders are not a homogenous collection of nonscientists. You should approach each group as if you were traveling to another country. What are their values? What language do they speak? How do they define success? What do they want from you?

Network is the key to delivering actionable science. As former Coast Guard commandant Thad Allen liked to say, "You don't exchange business cards during a crisis." You may be an expert in your field, but it's unlikely a responder is going to come to you—unless you already know each other. And if you want that relationship to be a mutually beneficial one, you will need to establish trust and understanding ahead of time. In many cases, the same is true of journalists. Sometimes a media network or reporter will actually try to tell you what to say in order to support their story. This is much less likely to happen if you have already established yourself as a trusted source.

How do we, as scientists, weigh the importance of what we say publicly versus the real-world consequences of such statements? Trying to gauge these consequences is the challenge posed when scientists *speak out*. While scientists have a long history of publicly challenging corporate narratives (e.g., the detrimental effects of tobacco and lead on human health), creating background noise or delivering noninformation—speculations with no supporting data—can wind up doing more harm than good.

Rooted in career challenges, *competition* between fellow scientists is real. While the benefits of such competition are repeatedly touted, too often the negative impacts of the winner-takes-all model are swept aside. Research might be duplicated, secrecy can take precedence over collaboration, and some scientists even burn out or become disenchanted with the entire process, dropping out of the field altogether. In a crisis, this competition can be exacerbated by the fog of war. When you're in the heat of the moment—trying to make the next big breakthrough, or get a study published in *Science* before a rival—it can be easy to lose sight of the bigger picture.

Communicating science is difficult because of the *process*, which, as we all know, is iterative by nature. It is a jigsaw puzzle, and each discovery is simply another piece that gradually reveals part of the larger picture. However, in a crisis, this key element—the hindsight and self-correcting course afforded by the passage of years, decades, and centuries—does not exist. How can scientists effectively communicate when the puzzle they are working on has only a few interconnected pieces and the clock is ticking?

The onset of a crisis event is often marked by high levels of uncertainty and anxiety, as the public is hungry for information that scientists are unable to provide. The resulting information vacuum is thus the perfect environment for the growth of conspiracy theories, whose ready-made answers cause real harm to society. While social media is often tied to the growth of such of *misinformation*, platforms like Twitter and TikTok can be a force of good too. From helping you improve fundamental storytelling skills to serving as a convenient place to share facts and data, this is one way to combat the rise of deliberate misinformation. However, it's important to remember that more than anything, empathy—not know-it-all elitism—is the key to getting the public to trust you.

Although the 1966 Freedom of Information Act was passed to increase transparency and accountability in the government, it has also been turned on its head to harass scientists at public institutions, by allowing industry and antiscience activists to demand every scrap of information about a research project, from marked-up rough drafts and handwritten notes to private telephone records. These sorts of *legal* tactics are designed to go beyond simply gaining information needed for reproducibility—they aim to put the maximum amount of psychological and financial pressure on the individuals involved.

Making a positive *impact* is our ultimate goal. Learning how to communicate begins with the people you are closest to: your family and friends. If you can't explain your research to your parents or keep your neighbor engaged at a backyard barbecue, how will you communicate your bottom line in a media interview or speak succinctly during a congressional hearing? Your goal should be to make every interaction so enriching and fruitful that people talk about the experience over dinner in a positive manner. You never want to miss an opportunity.

The last challenge is *teamwork*: in order to deliver content in a meaningful way, scientists have to partner with professionals outside their field. We need to overcome our bias that you need a PhD and published papers to be an expert. Being a team player entails building trust with the people you are working with, whether they are graphic artists, your institution's PR team, or government responders who don't run their own labs. Scientists will rarely save the day single-handedly.

Junior scientists are often the most excited when it comes to venturing outside the Ivory Tower. They want to make a difference, but first they have to build their reputations as scientists and figure out how the world works. I finish with practical advice on how to gain experience and develop connections with the world outside the research lab, with a particular focus on patience, starting at the local community level, and focusing on low-risk, high-reward outcomes.

This brings me to my final point. Some scientists, particularly those in the first group, who don't return calls or emails, tend to view outreach as a chore. I argue that communication has another aspect to it, one that is ultimately more fulfilling: it can make you a better person and a better scientist. Many of us suffer from impostor syndrome. We have moments where we question our own self-worth and the value of our research. But in communicating your science with the outside world, you have the opportunity to feel good about yourself. To become a better person.

A crisis can be a disaster, but it can also be an opportunity. An opportunity for you to make a difference. Why would you pass that by?

Notes

1 www.merriam-webster.com/dictionary/crisis.
2 K. Langin, "In a first, U.S. private sector employs nearly as many Ph.D.s as schools do," *Science*, March 12, 2019, www.science.org/careers/2019/03/first-us-private-sector-employs-nearly-many-phds-schools-do.

Part I
The Characters

1   Our Hero, the Scientist

Let's begin by talking about something decidedly unscientific: the hero archetype. Because who doesn't want to be a hero? It doesn't matter if we're talking about our personal or professional lives, we all want to be someone who successfully navigates and overcomes whatever challenges might come our way.

Now, maybe I am going against the pop culture narrative when I say: scientists are already heroes. We all know first-hand the countless hours we spend at the lab or in the field, our tireless devotion to advancing an understanding of the material world, the innumerable discoveries and breakthroughs that have radically changed people's lives for the better over the course of millennia.

Unfortunately, sometimes it feels like nobody else really cares. This was my experience in late 2009, when after ten years of doing groundbreaking studies on the chemistry of oil spills, I was so frustrated that I wasn't making a difference—that the responders who contained and cleaned up oil spills didn't value my work—that I was ready to give it up and go into an entirely new research field.

And then on April 20, 2010, a few months after beginning this transition, an oil rig in the Gulf of Mexico exploded.

Two days after the explosion, my friend Rob Ricker, who at the time was working for the National Oceanic and Atmospheric Administration (NOAA), gave me a call at MIT, where I was about to give a talk on communicating science.

"Hey Chris, did you see the news about the rig?"
"Yeah," I said. "That was heartbreaking."
There was a pause.
"Listen, I think we're going to need your help."

At the time, all I knew was that eleven workers had died in an explosion, and that the rig, *Deepwater Horizon*, was still on fire. I thought the explosion was linked to the fuel used to power the drilling rig. While 800,000 gallons of diesel fuel may sound like a big number, the location of the rig and other circumstances made it seem, in terms of contaminants, like it wasn't that big a deal. And so I declined.

DOI: 10.4324/9781003341871-3

"No thanks," I said. "I'm out. I've gone into plastics."

And that, I thought, was that.

A few days later, what was actually happening on the ocean floor still hadn't gone public. I received another call, this time from my good friend Bob Haddad, from NOAA's Office of Response and Restoration and one of the U.S. Government's lead scientists for the unfolding crisis. Bob, apparently, knew something that I didn't.

"Quit your whining," he said. "You *will* get involved. You *will* make a difference. This spill is going to change your life. We need you."

Let's press pause here. I had just begun transitioning my research into a new field precisely because I felt I wasn't making a difference. Did I really want to set myself up for disappointment yet again? At the same time, I knew that if this was as serious as Bob was making it out to be, I was ideally positioned: I was already familiar with how the industry worked; I had already done dozens of interviews about oil spills with the media—and learned from all the mistakes I had made; I had already majorly pissed off—and made up with—key responders, in the process coming to appreciate their own values and needs; I had lectured and collaborated with oil spill experts at the United States Coast Guard Academy; I had received communication training from Jane Lubchenco, then head of NOAA; and I even walked my dog with John Holdren, President Obama's science czar, who happened to live in the same town as I did.

Bob and I discussed the options. Could I work as a government consultant and an independent scientist at the same time? He promised to get back to me.

The following day it was revealed that an estimated one thousand barrels of oil per day were leaking from BP's well on the sea floor, nearly one mile beneath the ocean surface. The Coast Guard only worked with surface spills—a deep-water rupture was aqua incognita. For everyone. Well, almost everyone. Except for marine scientists.

Soon after this, Bob called me back and gave me the green light to conduct my own research and talk to the press, so long as I remained available to brief NOAA around the clock. By the end of April, I had given fifteen interviews to the media. I was not the only scientist whose phone was ringing off the hook.

We were finally going to get a chance. A chance to make a difference on the frontlines of a crisis.

Fast-forward to mid-August 2010, a month after the well had finally been sealed. By this point, independent scientists had made a number of substantial contributions, including accurately measuring how much oil had been released into the Gulf, mapping a twenty-two-mile-long subsurface plume of hydrocarbons, and determining how the currents were going to distribute the polluted water. Initial containment, damage assessment, and environmental restoration work all hinged on our contributions.

And yet, both the government and the industry hated us more than ever. The public had soured on standard science-based cleanup efforts, believing they were making a bad situation worse—an eerie foreshadowing of vaccination hesitancy during the COVID-19 pandemic. And the media was having a field day, ferreting out petty personal disputes between researchers and cherry-picking the most sensational end-of-world claims made by anyone with a PhD attached to their name. I got into an argument with a reporter during a taping of *Good Morning America*, which ended with him pulling the plug and storming off.

This wasn't how the hero's journey was supposed to end. Somehow, even in victory, it all felt like a net negative.

So what went wrong?

The Hero, Flaws and All

High school English flashback. In order for the hero to finally defeat the villain, first they have to overcome their own worst enemy: themselves. Let's take a moment and do some serious soul searching. Just what, exactly, is a scientist's Achilles' heel? Could it be … maybe … communication?

Okay, this is hardly a secret. There are at least a dozen books—good ones, too—that journalists have written with the aim of helping scientists communicate their research and knowledge to the general public, via the media. But the core problem is larger than mastering the art of the soundbite, or remembering to use the imperial system instead of metric, or to avoid speaking Romulan when negotiating with Klingons, or even to avoid nerdy Star Trek references—though, of course, this is all good advice (except for not using Star Trek references).

The core problem is rooted in academia itself: communication with nonscientists is not appreciated. Until science culture changes—along with outsiders' perceptions of it—scientists will continue to be poor communicators.

Why is this?

Because we all know that talking to the Other—people who are not scientists—is inherently risky. Very risky. Your words and data can be weaponized, misinterpreted, misrepresented. Your colleagues might think you're a sellout because you get too much attention—look no further than Carl Sagan, who was denied tenure at Harvard (but later validated) and never admitted to the National Academy of Sciences, allegedly because his efforts to communicate science to a wider audience were a bit *too* successful. Or maybe you'll be labeled an advocate: someone who uses their scientific training and knowledge in order to achieve a certain goal. You might make a mistake or find yourself in the middle of a public feud. Mistakes for scientists are sticky. They don't go away the next day, or the next year, or even five years after that. They could completely derail your career.

In short, there are plenty of reasons not to engage with nonscientists.

"Wait," I hear you say. "What about outreach?"

Good point. In fact, this is a great place to start. Everyone who has ever applied for funding, or for a promotion, knows they have to consider outreach. *How does your research impact society?* On its own, this is the million-dollar question. Who doesn't want to make a difference?

And yet, too often, scientists view this step as a chore, rather than looking at it as a chance to improve themselves and their science. Because outreach has no universal currency, there is no clear incentive to engage with the public in the current academic model. And this is a problem.

Box 1.1 From Carl Sagan to Hakeem M. Oluseyi: How Scientific Outreach Has Changed Over the Past Fifty Years

Carl Sagan's legacy as one of the twentieth century's greatest science communicators is unquestioned. But despite—or, perhaps, because of—his unmatched success, he seemed to rub many of his fellow scientists the wrong way. When he came of age in the 1960s, he made a lot of his colleagues nervous. He was excited, passionate, and, worst of all, intent on communicating outside of peer-reviewed literature.

In many ways, Sagan still embodies many scientists' deeply rooted fears: if you stray too far from the rigid confines of academia, you will be punished—or "Saganized." Everyone knows the price he paid for popularizing science: denied tenure at Harvard, shunned by some of his colleagues, and later snubbed by the National Academy of Sciences.

Have times changed? Well, it depends on who you ask. But there are increasing numbers of scientists out there who, like Sagan in his day, are continuing to blaze the path forward. One such individual is Hakeem M. Oluseyi. A Stanford-trained astrophysicist, Oluseyi has one of the most unconventional scientific backgrounds you'll ever read about.[1] He also has one of the more varied CVs you'll see from a researcher: he's dabbled in everything from chief science officer at Discovery Communications to "bakineering" judge on a Netflix cooking show. And he has eleven patents.

But it's not just in the United States where Oluseyi has had an impact. He is equally committed to outreach in developing countries. As he stated during an interview with TED:

> Given my life experience, I recognize the impact of culture and identity on the choices that a person makes. If you put a [research-grade] telescope in a country and you create an educational outreach program around it, you're going to get the kid here and there who says, "Hey, I want to be involved in this," and they'll spend their entire summer doing observations and taking data and analyzing data. ... There's a story of Africa, a perception that it's a place without science or scientists. And it's a self-fulfilling prophecy when you don't enable them to do real science.[2]

Risk and reward: you can't have one without the other. Learning how to manage career risk is the first step in successful communication. But you can't let it be a barrier that keeps you from inspiring others to get excited about science. For that brings a reward like no other.

The *North Cape* Oil Spill

In January 1996, when I was a student at the Graduate School of Oceanography at the University of Rhode Island, there happened to be a significant oil spill not fifteen miles from campus. At the time, I was studying the impact on coastal waters when you build roads from recycled car tires. But prior to graduate school, I had worked at two jobs that examined aspects of oil spills. So I knew a little bit about them. And naturally, from a scientific perspective, I was intrigued.

However, right as the spill took place, my advisor, Professor James Quinn, was about to leave on a two-week vacation. Before he left, he turned to me and said,

> Whatever you do, do not get involved. If a journalist finds out you're collecting samples and approaches you, do not say anything. It's very dangerous. You do not have the training. The media is looking to play *gotcha*, and if you make a mistake, you could ruin your career.

Now, my advisor knew me well. He knew I wouldn't be able to resist poking around. And he was right, because the second he was gone, I basically commandeered his lab and started taking samples and analyzing them.

This didn't bother him—what he was worried about was that I would go public and say something stupid. Because among scientists, there is little value associated with having a positive interaction with the media. Let's say I had spoken to a journalist and did a decent job. Maybe I would get one bonus point. But if I did a bad job—which was an entirely probable outcome—I would wind up with minus five points.

And he was absolutely right to tell me this. I was just a clueless, cocky kid. I would have been super excited to talk to the outside world about what I was doing. But I would have had absolutely no idea how to do it. I didn't understand the consequences. And my advisor did. And that's why I can still remember the look on his face when he said, "Whatever you do, *do not* talk to the media." He wasn't a threatening man at all. But his expression made it clear that this was off limits.

As it turns out, there are two sides to this story. About a year and a half later, the research that I conducted on the *North Cape* oil spill started attracting a lot of attention. The reason for this was that my studies revealed that the barge was carrying two types of oil, and not just one, as originally assumed. In addition to the diesel fuel that everyone knew had spilled, the *North Cape* also had a shipment of home heating oil. This was significant because that particular home heating oil was twice as toxic as diesel fuel.

RISKS   REWARDS

Figure 1.1 Communicating science is risky yet rewarding. It is important to identify
and weigh your own personal risks and rewards.

When the Rhode Island Department of Environmental Management confirmed my findings, this went on to impact the damage estimates and the
amount of money the state received to conduct restoration work. And when
the *Providence Journal* got wind that it was a native Rhode Islander who had
made the discovery, they naturally wanted to profile me. So in the end, this oil
spill wound up being my first interaction with the media after all. I had two
back-to-back interviews: the first was an above-the-fold piece that focused on
the science side of the story; the second was a human interest piece, focusing
on my local background and how I wound up giving back to my community.

And it was the first reporter, a well-known environmental journalist by the
name of Peter Lord, who really made a difference in my own career. The
thing was, I didn't even know it at the time. I was just excited to be a local
hero. I felt validated. It even carried me through a rough patch in my personal life. What I didn't realize was that, just as my advisor had predicted, I
made a lot of mistakes during that interview.

A decade later, I was coteaching a class on science communication at my
institution in Woods Hole. Part of this class included having a science journalist mentor students while they wrote an article. And one of the mentors

that I invited to the class was Peter Lord from the *Providence Journal*. On his first day in class, Peter stood up and said, "When you give an interview, don't do what Chris did."

"For example," he went on, "at one point I asked him: 'How did you discover that there were two different types of oil on that barge when a team of NOAA scientists never managed to figure this out?' And do you know what he said?" My students all leaned forward. Who doesn't enjoy seeing their teacher get roasted?

> *"I'm a better scientist."* I mean, this kid was twenty-something years old. He was still a graduate student. And he just publicly insulted a group of people he might wind up having to work with in the future. Now, this wasn't the story. So I didn't run that quote because I didn't want to hurt his career.

This particular anecdote illustrates two points: the first is that my advisor was right to discourage me from talking to the media. But his assessment of the dynamics at work was too conservative. The media is not fundamentally your enemy. How you are depicted in an interview depends a lot on the experience of the individual journalist and the angle of the piece they are writing. Understanding what to say and how to say it requires training—and the training that most graduate programs currently provide is insufficient. I'll come back to this point later.

The second point is that there is a lot of ego in our field. In itself, competition is not a bad thing. But an inability to read the room and personal disputes are certainly things that a journalist will pick up on. Not everybody will run with it. But eventually, making disparaging remarks about your colleagues will come up back to bite you. And more importantly, it only distracts from your main goal, which is communicating science.

Box 1.2 What the Experts Say: Media Coverage Increases Citations

In a landmark study, David Phillips and his colleagues showed how the *New York Times*'s coverage of articles in the *New England Journal of Medicine* (*NEJM*) led to increased citations in the first ten years of publication. Specifically, the authors tracked twenty-five *NEJM* publications covered by the *Times* and thirty-three *NEJM* publications from the same issue that were not covered (and which served as a control). The findings revealed, in the ten years that followed, researchers cited publications covered in the lay press at a significantly greater rate than those in the control group.

Phillips and his fellow researchers also took advantage of a twelve-week-long strike in 1978 when the *Times* printed an "official record" but did not actually sell any copies. During that period, the paper covered nine articles published in the *NEJM*. When compared with sixteen controls over the same twelve-week period, there was no difference in number of citations for either group. Phillips concluded that "coverage of medical research in the popular press amplifies the transmission of medical information from the scientific literature to the research community."[3]

In a more recent follow-up, Parker Sage Anderson examined 818 peer-reviewed publications in five journals in the fields of sport science, exercise, and physiology and reached the same conclusion.[4] So what's the takeaway? Contrary to the traditional view that interviews with the media carry more risk than reward, studies have proven that coverage in the lay media leads to increased citations, and thus a higher *h*-index for researchers.

So how do we overcome our flaws? How do we avoid winding up like Achilles—the Greek army's greatest warrior—whose excessive pride not only led to the death of his closest friend and lover, but also to his own downfall on a Trojan battlefield? Like many things in life, there is no one answer. There are several answers, and we will continue to explore them throughout this book.

To be sure, the failure to communicate science effectively is larger than scientists themselves. One only needs to look at the interplay of different stakeholder groups during the COVID-19 pandemic for proof of just how complex, and interdependent, this task has become—not only in North America, but around the world.

But one thing is certain. If we don't recognize and address our own flaws—both as individuals and as part of the larger scientific community—communicating science isn't going to get any easier.

Notes

1 See his memoir, *A Quantum Life: My Unlikely Journey from the Street to the Stars* (NY: Ballantine, 2021).
2 Karen Frances Eng, "Rise of a Gangsta Nerd: Fellows Friday with Hakeem Oluseyi," *TED blog*, October 5, 2012.
3 D. P. Phillips et al., "Importance of the Lay Press in the Transmission of Medical Knowledge to the Scientific Community," *New England Journal of Medicine* 325 (October 17, 1991): https://doi.org/10.1056/NEJM199110173251620.
4 P. S. Anderson et al., "A Case Study Exploring Associations between Popular Media Attention of Scientific Research and Scientific Citations," *PLoS ONE* 15, no. 7 (2015): https://doi.org/10.1371/journal.pone.0234912.

2 The Supporting Cast

Villain ... or Partner?

During *Deepwater Horizon*, some scientists seemed to forget that the enemy wasn't BP or the government. It was oil. Over 1,300 miles of shoreline along five coastal states was contaminated. Every part of the marine ecosystem—from plankton and invertebrates to fish and birds to sea mammals—was affected. Responders were doing everything in their power to limit the damage. So why did it seem like some scientists and journalists were intent on making their job harder still? And why didn't the government set up a dedicated channel for scientists to communicate their findings to responders?

During a crisis, different stakeholder groups find themselves thrust together in relationships that seem to be cobbled together on the fly. Whether or not these relationships are functional depends on just how well the two sides understand each other.

It is important to remember: nonscientists are not a homogeneous group. This is not us versus them. Each stakeholder has its own distinct culture, language, and value system, and learning how to communicate with each group requires training, experience, and mutual trust.

The Partner

In many stories, the main character has a close friend, partner, or love interest. Someone they can rely on. Someone who makes them better. And no protagonist will be able to overcome the villain in their story without the help of this person. In an ideal world, the different stakeholder groups that scientists work with during a crisis should be seen as partners, not antagonists. Take, for example, Peter Lord, the *Providence Journal* science reporter I mentioned in the first chapter. When I made a litany of mistakes during my first ever interview, Peter covered for me. He made me look good, because he understood my culture, even though I didn't understand his. And even though we spoke different languages, we were both committed to the same outcome: making my research known to the public. Because this was something that directly affected the lives of everyone in our community.

But what if the *Providence Journal* had been short on reporters that day and sent a sports journalist to interview me instead? Would that person have

DOI: 10.4324/9781003341871-4

understood the consequences of running a quote that made government scientists look bad? Maybe—but maybe not. And if not, then my potential partner would have become the story's villain, just as my advisor had warned.

The key to a successful relationship depends on knowing how these different stakeholder groups define success, and how you can help—or hinder—them. Once you understand this, knowing what to say, and how to say it, will become considerably easier.

To use an analogy: let's say you have a restaurant. And you're really good at making lasagna. Would you serve the same plate of lasagna to all your customers, every single day? It doesn't matter how talented you are in the kitchen, this type of thinking would doom your business immediately. Wouldn't it make more sense to offer a menu, so that your customers could choose the type of dish that was most suited to their tastes? Some people just want a simple hot dog and a Coke, served in under five minutes to eat on the go. Others want a five-course meal, served over the duration of an entire evening, and expect the restaurant to look the part. And yet, I continue to see people consistently ignore the menu—they serve the same message, in the same way, to everyone they meet. And not just among scientists. So who are these different groups, and how can we learn what they want?

Responders

Goal: To contain a crisis
Challenge: Different time frame; no official communication channel
Needs: Actionable science as quickly as possible

Responders are the group of people on the frontlines of a crisis: those who keep a bad situation from becoming worse. They might include members of the Coast Guard, the National Ocean and Atmospheric Administration (NOAA), firefighters, federal and state environmental agencies, the police, the military, industry leaders, and, in the case of a pandemic, health care professionals.

A responder's job is only done when there is nothing left that humans can do to make a difference. Responders do use scientific data, but they often don't like dealing with independent scientists. Why? Because they need information fast. They don't have as high a bar as you do when it comes to accuracy and certainty. They are more willing to work with what they can get. They may not fund you. They don't like to gamble and use unproven technology. And they definitely don't care about peer-reviewed papers in the heat of the moment. Many of them do publish papers down the road, but it is not their driving motivation.

This doesn't mean that responders don't value or need science—they do. But every day that goes by is another 25,000 acres of forest that has burned down, or another fifty miles of coastline that has been contaminated. They simply don't have the time to sit around and wait for perfect data.

During *Deepwater Horizon*, responders had three primary goals: to contain the contaminated water (using chemical dispersants, controlled burns, booms, and skimming); to stop the flow of oil emanating from the sea floor; and to assess the environmental impacts of the spill and help keep the fishing and tourist industries open when possible. You could also argue they had a fourth goal: to control the narrative in the press.

As it turned out, NOAA and the Coast Guard did work with several teams of scientists to attain some of these goals—we'll get to this later. But there were many other scientists in the Gulf who did not have access to the responders, and felt frustrated because of this. The government didn't even invite our community to the table until May 21, a full month after the explosion. That event, a hastily organized conference at the EPA head-quarters in DC, felt half-hearted. The agency's administrator, Lisa Jackson, did no more than pop her head into the conference room.

The end result was that many scientists felt compelled to share their findings—and sometimes their opinions as well—via the media. This only wound up irri-tating the response community, because they suddenly found themselves having to take time to field questions they didn't know the answers to, instead of being able to focus on doing their job.

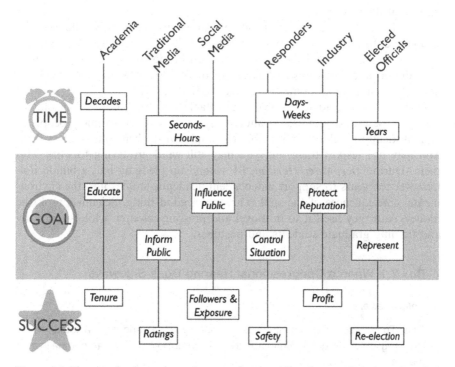

Figure 2.1 Communicating science is more than avoiding jargon. It is knowing that your audience is heterogeneous and appreciating their value systems.

Industry

Goal: To be profitable, to maintain a positive public image
Challenge: May seek to misrepresent your data to reduce their own financial liability; may refuse to share their own data; capable of waging long legal battles; must answer to their shareholders
Needs: Actionable science that helps them solve the problem

While industry is not always involved in crisis events, in the cases where they are, they will play a critically important role. During *Deepwater Horizon*, critics of the response often wondered out loud: Who is in charge of the cleanup effort? BP or the federal government? In fact, BP *was* in charge of the operations—this is federal law. But because the logistics of an environmental response are poorly understood, to most people it seemed as if the Coast Guard was doing no more than enforcing the industry's directives (see Chapter 3 for more on this topic).

Bear with me for a moment, because I am going to say something that may be tough to swallow: industry is not always the bad guy. If you maintain a mindset that their goal is to destroy the planet, then cooperation will be difficult. But it is important to remember that a company's goal is not to be good, evil, conservative, or liberal. It is to make money. Full stop. This is an important distinction.

And the industry has a *lot* of money. Politicians are heavily influenced by lobbying dollars and campaign contributions. The public depends on them for employment. Even amid the recent downturn in the energy sector, in Louisiana, the oil and gas industry accounted for 249,800 jobs and 26% ($73 billion) of the state gross domestic product (GDP) in 2019.[1]

Does industry need to be held accountable for their actions? Absolutely. Yet, because industry money is intertwined in the lives of so many people, it is, realistically speaking, impossible to simply write them off as the villain. You may not respect them, but you may still wind up having to work with them. During *Deepwater Horizon*, BP wound up pledging half a billion dollars over ten years to support independent academic studies into the environmental consequences of the spill. This was a good thing. Scientists took the money. And they used it to do novel and pressing research while supporting and training graduate students and postdocs.

Box 2.1 When a Congressional Hearing Goes Sideways

Policymakers

Goal: To represent their constituents, provide oversight, and get reelected
Challenge: You've got five minutes to make your point in a clear, compelling manner
Needs: Expert opinions on policy matters

During a major crisis, politicians want to get involved, so they hold hearings and shine a light on what's happening for their constituents. The results of these hearings could eventually influence policy decisions down the road, but it's less likely they'll affect the outcome of the crisis itself.

Over the course of my career, I've been asked to testify five times before Congress, and once before a Presidential Commission, in addition to other appearances at the state level in Massachusetts and Rhode Island. Hearings ranged from investigating the toxicity of Corexit, one of the dispersants used to clean up oil spills, to whether or not NOAA has enough scientific funding.

Fitting in to congressional culture is fairly straightforward: always defer to the people holding the hearing. Don't go toe to toe with them, because you will never win. Be respectful, professional, and follow the dress code: men should wear a suit and tie, and women should wear either a dress that covers the shoulders or a jacket and slacks.

While this might seem like common sense, not everyone has received the memo. During one particular testimony, I was watching two other scientists from the sidelines. The first guy showed up wearing tight designer jeans, Italian sneakers, and a leather jacket. Not only did he insult his audience by not dressing in the appropriate way, he obviously hadn't prepared either, because he didn't give a very good statement.

The next scientist was polite and did wear a suit, but, unfortunately, had decided to use PowerPoint for his presentation. No one had told him not to do this. The lights remained on and he had no screen on which to display his presentation, so he wound up spending the majority of his allotted five minutes fiddling with a slide show that was projected onto a painted wall with a bit of American flag in the corner. And then, to top it off, he began his speech by telling the committee something along the lines of: "I'm going to go through a little bit of math, but I'll slow it down for you."

Those two scientists broke basic rules in their testimonies: the first was inappropriately dressed and conveyed an air of not taking the hearing seriously. The second wasted most of his time fiddling with a projector, and then when he finally did say something, spoke to the committee members as if they were his students. Though neither of them appeared to have any training on how to speak to Congress, they hadn't thought to ask for advice. And because of that, we all missed an opportunity.

Congressional committees want to look like they're doing something useful (and hopefully they are), but on that particular day they didn't get the answers they were looking for. It was a bad day for science. After the hearing was over, I left the Capitol, full of remorse. The government had asked academia to testify, and we hadn't even made the effort to respect our audience's culture, let alone give a coherent presentation.

The Media

Goal: To provide comprehensive coverage of a crisis; to ensure there is accountability; to attract readers, viewers, and listeners

Challenge: Different time frame; some reporters will sensationalize or misrepresent scientific data

Needs: The bottom line; a good quote that supports the angle of their story

This is the most diverse and complicated stakeholder group. Journalists produce newspapers, magazines, websites, radio shows, podcasts, and television programs. Everyone needs journalists. At the same time, everyone loves to hate them too. Regardless of a journalist's medium, most are deadline driven. And most adhere to the belief that a good story must have a beginning, a middle, and an end.

These last two points are the crux of the problem when it comes to communicating your research in the media. The timescale reporters live in consists of minutes and hours. Scientists, in contrast, live in a timescale that consists of months and years. Media outlets often focus on newsworthy one-off events, while scientists are engaged in an incremental process that has no identifiable endpoint.

In a crisis, these differences are accentuated. In a crisis as big as *Deepwater Horizon*, they are further complicated by the fact that every news outlet in the country is covering the same event. How do all these journalists distinguish themselves from one another? Good reporters will find a unique angle. Lazy reporters, or reporters with their own agenda, will sensationalize or misrepresent.

In fact, one thing scientists have a hard time understanding about journalists is that someone who has no knowledge about a subject may be assigned to write about it in under eight hours. This doesn't mean it's not possible, because it happens all the time. But it can be a potentially risky situation for a scientist who is worried about misrepresentation, or having their research become weaponized.

There are also a select group of journalists who specialize in science and the environment. These journalists have been to labs, understand their subject matter, and may even have a background in science themselves. But they are the minority, and in order to work with them, you will need to establish a relationship ahead of time. Establishing trust and a mutually beneficial relationship is the key to success with any stakeholder group.

However, the traditional media are no longer the only ones who control the narrative—social media has forever changed the nature and flow of information—and misinformation—around the world.

The Affected Public

Goal: To find out when they can resume everyday life

Challenge: May get caught in the cross-fire of public speculation; may not want to hear what science has to say

Needs: The facts

Just like we often use the term *nonscientists*, implying that these people are a homogeneous, unified group, so too do we talk about the *public*. However, the public is no more homogeneous than are nonscientists. Personally, I divide the public into three main groups: the affected public (those directly affected by an event), the contentious public (those who dispute the mainstream narrative), and the concerned public (those who are following the events but are not directly affected).

Scientists should always consider how the affected public might interpret or use their statements. Some of the hardest conversations I've had over the course of my career have been with people who are directly affected by a crisis. The affected public are desperate for information—real information, because their lives are on the line.

And when communicating with this group, the worst thing you can do as a scientist is to provide noninformation. That is, to make a speculation with no supporting data. Saying something like, "The Gulf will be dead for one hundred years after this spill," or "Half a dozen species of fish will go extinct because of this" during an oil spill is counterproductive.

What were scientists thinking when they made such statements? They were probably trying to challenge the official BP narrative during the first month: *We've got it all under control. Everything's fine.* In the moment, they thought they were doing the right thing.

But when residents hear catastrophic statements like this, that's when the local housing market bottoms out, when domestic violence and suicide increase, and when people sell their fishing boats. They might think their livelihoods are ruined forever. Words have consequences. Before you make a public speculation or sweeping statement, always consider how the people who are directly affected by the event might interpret what you say.

The Contentious Public

Goal: To dispute the mainstream narrative
Challenge: May actively try to refute or misrepresent your findings
Needs: To be heard

The contentious public are those people who are looking to get into the mix—they might be trying to blow the negative aspects of a situation out of proportion, or perhaps they're convinced that the government or industry is involved in a large-scale cover-up or conspiracy. They might quite simply be antiscience.

If some of the hardest conversations I've had as a scientist have been with members of the affected public, some of the most challenging conversations have been with the contentious public. How do you engage with someone who doesn't believe you? Who thinks you are a corporate or government shill? Who dismisses you as an elitist?

As scientists and health professionals found out during COVID-19, there is no easy solution to dealing with the contentious public. And the arguments that scientists make—based on data and rationality—may even make things worse. Rolling out your academic credentials to establish your expertise? Also a no-no. If we imagine the relationship between scientists and the contentious public to be an acrimonious marriage, we might say that the key to resolving the tension is not about establishing who is right. It is about listening to what the other person has to say and validating their feelings. Unfortunately, making use of this tactic in the heat of the moment is easier said than done.

You may be tempted to simply ignore this group. Or you may lose patience and waste a lot of time and energy debating them. And yet, as we have seen, this is a group that cannot be wished or rationalized away. Because of social media, the influence of the contentious public has continued to grow in a manner disproportionate to their size. We'll talk more about them, and scientific misinformation in general, in Chapter 7.

Box 2.2 Doctor Fauci

Science communication is inherently challenging, as anyone who has stood before an audience of laypeople can attest. And in the twenty-first century, no other American scientist has done this more, and in such a high-stakes environment, as Dr. Anthony Fauci.

Appointed the director of the National Institute of Allergy and Infectious Diseases (NIAID) in 1984, Dr. Fauci briefed six US presidents before becoming the face of the COVID-19 pandemic response in 2020. There could not have been a better-suited or more experienced scientist in the world to lead the charge. Even while the administration faltered on delivering a clear, consistent message, Dr. Fauci went above and beyond in his attempt to communicate the facts. Not only did he testify before Congress and regularly brief former president Trump, but he also participated in regular press conferences and even took to social media in an attempt to reach increasingly greater numbers of people.

His dedication to sharing critical information about the disease across multiple channels showcased his grasp of one of the key tenets of communication: know your audience. On one day, Americans might see Dr. Fauci appear before the Senate Health, Education, Labor, and Pensions Subcommittee. In this situation, he appeared somber and dressed the part, wearing a dark suit and tie (a uniform appropriate to the place and occasion) while describing the complicated, time-consuming process of developing and evaluating new vaccines. Despite the complexity of the subject, he used language that any Senator could understand—and most importantly, repeat—when asked why it was taking so long for the government to act: "We need to make sure it's safe, and we need to make sure it works."

On another day, we might see him chatting in a relaxed manner with NBA star Steph Curry on the latter's Instagram feed. What could have been dismissed as a publicity stunt by a sports icon turned into a remarkable display of how to reach people who might not otherwise have heard Fauci's message. This initial social media success led to further interviews and myth-busting sessions with talk show host Trevor Noah of *The Daily Show* and YouTuber Lilly Singh.

And one can only imagine the conversations with former president Trump, who built his brand on creating chaos and firing his employees. The fact that Fauci even managed to keep his job indicates that he also knew how to communicate with one of the more prickly politicians the United States has seen in recent history.

No matter whom he was speaking with, Dr. Fauci kept his strategy simple. Each interview was based on four things: here's what we know, here's what the models are showing, here's what we don't know, and here's what we need to do. As Dr. Fauci prepared to retire in December 2022, he reminded the next generation of scientists,

> It is our collective responsibility to ensure that public health policy decisions are driven by the best available data. Scientists and health workers can do their part by speaking up, including to new and old media sources, to share and explain in plain language the latest scientific findings as well as what remains to be learned.[2]

While scientists might respect Fauci for his credentials—his *h*-index is an astounding 227, meaning he is one of the most cited researchers in the world, largely because of his work on HIV immunology—his message resonated with his audiences for another reason: he displayed empathy with the affected public and he wasn't afraid to say, "I don't know." This tell-it-to-me-straight attitude helped win credibility with many Americans, even if members of the contentious public continued to try to discredit him.

But even as COVID-19 became increasingly partisan, Fauci made sure to keep his opinions to himself—he focused solely on delivering scientific content to a variety of stakeholders, because his main goal was to curb the worst of the pandemic's effects.[3]

The Concerned Public

Goal: To learn more about the crisis
Challenge: Getting your message across in a clear, effective manner
Needs: More information

The concerned, or curious, public is the easiest group to communicate with. They have their own opinions on the crisis, but their primary goal is to stay informed. They may be indirectly affected by events, but they are not in immediate danger of losing their house or job. And if they recognize you as an expert, then they will likely listen to what you have to say.

When talking with the concerned public, whether you're in line at the grocery store or being interviewed for a newspaper article, you simply have to aim to make every interaction a positive one, to communicate only what you know to be true, and to speak in a clear and engaging manner.

Box 2.3 What the Experts Say: Credibility and Previous Experience Are Key Determiners for Effective Communication

At the outset of COVID-19, members of the public, the media, and the government entrusted physicians and scientists to explain what the virus was, how to combat it, and how vaccines might help. Thrust into the spotlight, many physician-researchers, like Anthony Fauci and Chris Whitty (UK), became household names.[4] Polling and media coverage indicated that their audiences believed them to be "credible." In addition to being successful researchers, each had previous experience communicating science during a controversial crisis. This prepared them for the onslaught of media attention and enabled them to properly evaluate the risks and rewards associated with it.

Following the 1986 Chernobyl nuclear reactor disaster in the former Soviet Union, Hans Peter Peters and his colleagues surveyed West Germans on the credibility of scientific sources.[5] They concluded that credibility was driven by "expertness" and "trustworthiness," which was measured by the capacity to voice information accurately and clearly, respectively.

Fernando Simón (Spain) was praised for his "clear, concise explanations."[6] One of Chris Whitty's colleagues said, "People trust him because he bases his decisions on evidence."[7] Others have written that Salim Abdool Karim (South Africa) "stays in his lane and sticks to the evidence."[8] Jérôme Salomon (France) was recognized for his seriousness,[9] while Roberto Burioni's (Italy) "ill-tempered" and blunt approach endeared him to many, though some argued that he was too polarizing.[10]

Activists and Non-Governmental Organizations (NGOs)

Goal: To advance an agenda
Challenge: May try to misrepresent or cherry-pick your findings
Needs: Credibility from an independent source

NGOs can be tricky, because their goal is to advance an agenda. Sometimes that agenda is clear and driven by laudable motivations that you support. And sometimes it's not. Even if the pitch sounds good, you should always ask yourself, "Are they actually doing what they say they're doing?" The catch, of course, is that the average person does not have the time, or the resources, to investigate each and every NGO that reaches out to them.

For a scientist, this is complicated by the fact that our role is to provide unbiased, data-driven content. As Audrey Williams June wrote in *The Chronicle of Higher Education*, "Scholar-activists must be ready to fend off the perception that their activism taints their scholarship, or that they're going to indoctrinate students."[11] Ultimately, you may feel that the risks of crossing the line into activism are insignificant compared with the risks of not publicizing your research—scientists like Clair Patterson and James Hansen, for example, were willing to put their careers and reputations on the line for the sake of making an impact.[12] But, bear in mind, the risks are real.

While there are many excellent NGOs doing valuable work—for example, in my own career I have partnered several times with the Buzzards Bay Coalition, among others—there is always the possibility that an NGO will do whatever it takes to push their agenda. And this may include cherry-picking your findings or making inaccurate statements in order to fit their narrative.

If this happens and your colleagues see a quote from you, taken out of context, they might conclude that you've crossed the line and gone into policymaking instead of simply providing content. It's kind of like baseball: people go to the game to watch the players. No one ever talks about the umpires, *unless* they make a mistake—even if they called an otherwise perfect game. With NGOs and policymaking, scientists are the umpires. And you do not want to be remembered for having made a bad call, because then you will lose credibility within your own peer group.

Don't misunderstand: I am not criticizing the work of NGOs. But I am cognizant that there is a definite line between activism, which is goal-based, and independent science, which is results-based. Thus, even if I advise caution when speaking to NGOs that I am personally unfamiliar with, that doesn't mean I ignore them. If someone reaches out to me from an NGO, generally I'll reply to their request with an email and say something like, "You know what, I can't help you here. But this is a recent paper that has a lot of the answers you're looking for." Or I'll suggest they contact someone else who might be a better fit.

No matter what, whenever someone contacts me, I don't stick my head in the sand and pretend they don't exist. I try to leave them nourished. Even though a scientist's goal should never be to please other people, you can still find a way to make every encounter an enriching one.

Notes

1 ICF, "The Economic Impact of the Oil and Gas Industry in Louisiana," October 5, 2020, www.lmoga.com/assets/uploads/documents/LMOGA-ICF-Louisiana-Economic-Impact-Report-10.2020.pdf. However, some argue that the Gulf Coast oil industry is in decline and has already begun the inevitable shuttering of refineries and slashing of jobs as the United States transitions to renewable energy. See, e.g., Adam Mahoney, "Lessons from the Slow Death of Louisiana's Oil Industry," *Grist*, December 16,

2021, https://grist.org/climate-energy/it-doesnt-have-to-be-this-way-lessons-from-the-slow-death-of-louisianas-oil-industry.

2 A. Fauci, "A Message to the Next Generation of Scientists," *New York Times*, December 10, 2022.

3 I do believe that the NIH—or perhaps Dr. Fauci himself—did make a critical mistake during the pandemic: they did not appoint a successor or understudies to appear in public with him. Not only would it have been good communication training for the next generation, but having a deputy might have helped, if, for instance, he had gotten sick.

4 J. Henley, "Coronavirus: Meet the Scientists Who Are Now Household Names," *Guardian*, March 22, 2020, www.theguardian.com/world/2020/mar/22/corona virus-meet-the-scientists-who-are-now-household-names; M. Joubert, "From Top Scientist to Science Media Star during COVID-19: South Africa's Salim Abdool Karim," *South African Journal of Science* 116, no. 7/8 (2020): https://doi.org/10. 17159/sajs.2020/8450.

5 H. P. Peters, "The Credibility of Information Sources in West Germany after the Chernobyl Disaster," *Public Understanding of Science* 1, no. 3 (July 1992): https:// doi.org/10.1088/0963-6625/1/3/006.

6 C. Lopez, "Fernando Simón, el hombre que da la cara ante el coronavirus," *La Vanguardia*, March 7, 2020, www.lavanguardia.com/vida/20200307/473989540682/ fernando-simon-centro-emergencias-sanitarias-coronavirus.html.

7 I. Sample and H. Stewart, "'A Class Act': Chris Whitty, the Calm Authority Amid the Covid Crisis," *Guardian*, March 22, 2021, www.theguardian.com/world/2021/mar/22/a -class-act-sir-chris-whitty-the-calm-authority-amid-covid-crisis-chief-medical-officer.

8 M. Joubert, "From Top Scientist to Science Media Star."

9 J. Henley, "Coronavirus."

10 F. Brandmayr, "Public Epistemologies and Intellectual Interventions in Contemporary Italy," *International Journal of Politics, Culture, and Society* 34 (2021): https://doi.org/10.1007/s10767-019-09346-3.

11 B. Benderley, "The Value—and Risk—of Activism," *Science*, July 30, 2015. https:// doi.org/10.1126/science.caredit.a1500190.

12 On Patterson, see the Conclusion. For Hansen, see Elizabeth Kolbert, "The Climate Expert Who Delivered News That No One Wanted to Hear," *New Yorker*, June 29, 2009, www.newyorker.com/magazine/2009/06/29/the-catastrophist.

Part II
The Crisis

3 How a Crisis Response Unfolds
The Role of the Scientist

One of the defining moments of my early career was the *Bouchard 120* oil spill. Like the *North Cape* oil spill, this was a local event, which took place just twenty miles from my home. On April 27, 2003, the *Bouchard 120* tanker struck a rock ledge near Buzzards Bay, Massachusetts, spilling 98,000 gallons of bunker fuel just offshore. Because it was so close to Woods Hole, where I worked, my colleague Bob Nelson and I were able to collect samples the next morning and bring them back to my lab that same evening.

Although this was not the type of crisis that garnered much in the way of national attention, it was certainly a big deal in the local community. Part of the reason for this was that in 1969 another tanker, the *Florida*, had run aground just twenty miles from the *Bouchard 120*, which resulted in the environmental disaster known as Silent Autumn. An estimated 95% of all fish, shellfish, worms, and other invertebrates in one heavily impacted area in West Falmouth were killed within days.[1]

In 1999, I had studied the aftermath of the *Florida* spill and discovered that certain residual compounds from the No. 2 home heating oil were still present in the West Falmouth salt marshes. Scientifically, this was extremely interesting: why was nature able to break down most compounds, but not these particular ones? Why didn't microbes want to eat them?

This is the type of information that could be extremely valuable for someone developing next-generation biodegradable chemicals that might be released into the environment. A researcher would want to ensure that a new chemical doesn't contain any compounds or structural motifs that aren't palatable to microbes. After I published a paper on my study, it was picked up in the media, but with a considerably different angle: *the effects of oil spills continue to linger in marshlands for up to thirty years.*

The compounds that I had found existed only in one small area six inches deep in the mud—where they had an impact, it was very, very localized. The ecosystem as a whole was doing fine. What was interesting to me was not their impact, but their persistence. But this media angle, about oil spills lingering for decades after the fact, was still relatively fresh in people's minds when the *Bouchard* accident happened. And when I was invited to speak at a community meeting held at the New Bedford Whaling Museum six days after the *Bouchard*

DOI: 10.4324/9781003341871-6

spill, it was standing room only. The audience included members of every sta-keholder group: the affected public (including fishers), responders, industry reps, the media, local government officials, NGOs, and even a few fellow sci-entists like John Teal, who did groundbreaking work on the West Falmouth spill in 1969.

I'm not going to lie—I thought I did an amazing job that morning. I had really cool graphics that showed off the new gas chromatographic techniques Bob Nelson and I were using to perform our analyses. I told the crowd how studying this spill would change the field. I was dressed in a pressed blue suit. I was super excited about the science and considerably more enthusiastic than any other speaker there. Career-wise, it was good for me: I went on to publish five papers in the coming years, all based on my research, which cemented me as an oil spill scientist.

But after the meeting concluded, some members of the audience came up to me and said, "You know, I didn't understand a word you said, but I liked the way you said it. And those graphics were really cool!" I should have copped on immediately: *Chris, you didn't get an A+ on your presentation, you got an F.* But I was still too amped up to realize that the audience was hungry, and I hadn't nourished them. What most people wanted was comfort food: Are my kids going to get sick? Is it safe to go to the beach? Are all the lobsters going to die? When can we start fishing again? How many miles of coastline are going to be oiled? Instead of simply giving them a grilled cheese and a Coke, I had just ordered my audience a five-course meal. And they would have to wait several years before they could taste it.

Interestingly, the day before I gave this presentation, I got a call from an industry consultant.[2] She was brief and to the point:

> Chris, we saw you're one of the speakers tomorrow. We want to remind you that this is not going to be another West Falmouth [the location of the 1969 *Florida* spill]. The oil is not going to stick around for thirty years. It would be a big help if you didn't say that.

Okay … This sounded *exactly* like the sort of thing a powerful corporation would say to a whistleblower. I couldn't believe it. This woman didn't even have a lab. Who was she to tell me what I could or couldn't say about my own research? When I hung up the phone, I was so angry that I discounted her entirely. Her message failed to go through.

But here's the thing: she was right. In hindsight, I now realize that the first thing I should have told the audience that day was,

> I just wrote a paper on an oil spill that happened thirty years ago. Yes, there are still some compounds out there, but they are so incredibly trivial that it is not a big deal. Do not assume that this is the same scenario. There is going to be some damage to birds, little to no damage to salt

marshes, and I imagine that once the data comes back, all the fisheries are going to be open. Based on my past experiences and knowledge, this is not going to be the end of the world.

I could have talked about how the viscosity of the bunker fuel had changed over the past few days and how that might affect containment and cleanup. That was the type of information the audience was interested in. They didn't care about my breakthroughs using gas chromatography, apart from the fact that the slides looked cool.

As it turned out, there was another speaker that morning who also thought I missed the mark. This was Steve Lehmann, the National Oceanic and Atmospheric Administration's (NOAA) senior scientific support coordinator for the New England region. Steve was in charge of incorporating all the data and providing his personal insight in order to help the response team's incident commander make her decisions. When he got up on stage, he looked *terrible*. Like he hadn't slept for four nights straight (he hadn't). Like he was still dressed in the same old jeans and rumpled shirt that he had been wearing since the first day of the spill (he probably was).

Steve had already given multiple press conferences, and by this point just wanted to get back to doing his job rather than spend time talking to the public and the media. He hadn't even bothered to create a PowerPoint deck. Instead, he used an overhead projector! As I watched him give his presentation, all I could think to myself was, "No one is going to listen to a word he says. How did this guy get to be NOAA's scientific support coordinator? He doesn't even have a PhD!" Little did I know, this was not the last time I would cross paths with Steve Lehmann.

Shortly after the community meeting, I gave several interviews to the media and shared my findings. One thing I happened to mention was that the kind of bunker oil that spilled contains the compound naphthalene. This was just a statement of fact—I wasn't trying to suggest that this oil spill was akin to dumping 96,000 gallons of mothballs into the bay. But because I didn't explain myself clearly, several reporters went on to make an unintended connection between naphthalene, mothballs, and toxicity, and added to my credentials that I had just published a paper on how spilled oil could linger in salt marshes for up to thirty years.

You can see where this is going, right? The spill's environmental impacts were being spun in the wrong direction. When the response team read the news the next day, they weren't happy. In addition to containing a crisis, part of their job is to understand the optics of the situation and how the narrative is shaping up in the media. Not only do they need to contain the environmental impact of an event, they also need to contain the social impact. And a rogue scientist who had just stated that there was now naphthalene in Buzzards Bay? And the newspapers were claiming this particular compound was extremely toxic? Not good. When they gave their daily press conference, the media would want to know all about the naphthalene, and they would have to have the answers.

After learning of my findings, the spill's incident commander, Coast Guard captain Mary Landry, probably called Steve aside. She might have said something like this: "What's all this about naphthalene? Is it really going to kill all the marine life? Why wasn't I briefed about this? What are we going to tell the media?" Steve probably responded, "I'm not sure about the naphthalene, but this is bunker fuel. We have a playbook for bunker fuel, and I have thousands of hours of experience. We're on top of the situation."

Captain Landry shook her head. "Stop whatever you're doing and get me more info on naphthalene. We need to know how to respond to this."

Then Steve went back to his computer, sat down, and fired off an email to me. In it he wrote, "Lots of 'face time' on the news. Glad to see WHOI [Woods Hole Oceanographic Institution] getting the publicity. We need to talk more frequently. You had promised to share data and I have not seen much."

What is the role of a scientist in a crisis? Should we act as whistleblowers, whose job is to police the response efforts of the government and industry? Are we the experts that the media consult when they need a quote—only to find out later that the reporter has misinterpreted what we said? Or, just as frustrating, that the reporter already has an angle and he or she actually wants to tell us exactly what to say in order to support that angle? And how does a crisis response work, anyway?

In the aftermath of *Deepwater Horizon*, several notable leaders in the response community wondered out loud about these very questions. As Jane Lubchenco, the former head of NOAA, wrote in a paper published in 2012:

> Effective mechanisms are needed to enable rapid two-way communication with the broader scientific community. No single mechanism existed for us to communicate easily with the large, undefined, and interdisciplinary community of scientists. The US Government set up daily calls with governors, members of Congress from Gulf Coast states, parish presidents, and journalists. Unlike those easily identified groups, "scientists interested in the spill" were a challenging group to identify quickly and communicate with frequently and in the depth required for meaningful exchanges. ... For the most part, many scientists could get and share updates only through information in the public press, which led to considerable misunderstandings and great frustration.[3]

The crux of the relationship between the government and independent scientists is summarized perfectly in the final sentence: instead of speaking directly to one another, these two groups communicated through an intermediary. And not only was this intermediary sharing all the information publicly, it was also misrepresenting or sensationalizing certain claims.

The same thing that had happened during the *Bouchard 120* oil spill—when my inadvertent mention of naphthalene wound up interfering with the response team's ability to do their job—happened again in *Deepwater Horizon*. Only this time, it wasn't one scientist talking about a local spill. It was dozens of scientists talking about the largest environmental crisis in recent US history. The background noise was almost deafening.

Box 3.1 The Disobedience Award

Part of effective communication is recognizing the right time to challenge industry and government norms or narratives. In this chapter and the next, I talk a lot about the unintended consequences of delivering scientific non-information, and the detrimental effects this can have on the affected public and the response community.

However, do not mistake situational awareness for self-censorship. I am not saying you should keep your head below the parapet and avoid communicating your research. What I am saying is that you should be smart in how you announce your findings. Science has a long tradition of whistleblowers who have brought about important changes in environmental policies, despite facing intense pressure from industry and government, and we could all learn much from their courage and persistence.

One such person is Dr. JoAnn Burkholder, the director of the Center for Applied Aquatic Ecology at North Carolina State. In the early 1990s, Dr. Burkholder linked the dumping of agricultural nutrients into local watersheds to toxic algal blooms (two species of algae called *Pfiesteria*) that killed a million fish in North Carolina estuaries and the Chesapeake Bay, in addition to causing illness and severe cognitive impairment in humans exposed to the algae's neurotoxins. Thanks to her research, safety measures were put in place that temporarily banned fishing, closed beaches, and prevented swimming in the Neuse River in the years that followed. Her efforts were later documented in the book *And the Waters Turned to Blood*.

However, Dr. Burkholder was not without her critics—for over a decade, she had to endure personal attacks and attempts to discredit her research that stemmed from the agricultural industry, recalcitrant state health officials, and even fellow scientists. Those allied against her cast public skepticism on her lab work and scientific competence, and it was not until 2007, when a NOAA scientist identified a venom that the *Pfiesteria* microbe produces, that Dr. Burkholder's research was validated.

Another scientist who underwent public scrutiny and intimidation was Dr. Deborah Swackhamer (1954–2021), an environmental chemist who taught at the University of Minnesota. Dr. Swackhamer's early research focused on toxic contaminants such as polychlorinated biphenyl (PCB) and toxaphene and how they entered and moved through the environment in the Great Lakes region.

After Dr. Swackhamer's findings cast doubt on whether or not regional pulp and paper mills were creating toxaphene as a byproduct, which then was absorbed by the waters of Lake Michigan where it made its way into fish, the industry launched an intimidation campaign to try to get her to stop her research. As Dr. Swackhamer said, "They wanted to see records of my grants, my teaching materials, phone calls, all my data, for a thirteen-year period [using a request for public information]. We shipped off container after container of papers; they kept coming back for more and more information."[4]

Dr. Swackhamer later went on to serve on the Environmental Protection Agency's (EPA) advisory board from 2003 to 2012, and then acted as the chair of the EPA's Board of Scientific Counselors. In 2017, then EPA head Scott Pruitt's chief of staff tried to pressure her into changing a Congressional testimony that criticized the Trump administration: earlier in the year the White House had replaced half of the board's independent scientists with more industry-friendly appointees who lacked the proper credentials. She was demoted soon after, and when her term ended in 2018, she was not reappointed. Dr. Swackhamer was, however, nominated as finalist for MIT's Disobedience Award later that year.

The Incident Command System

In the United States, the government has a special protocol designed to incorporate multiple agencies into any emergency response, regardless of its nature. This is known as the Incident Command System (ICS). The origins of the ICS were developed in 1970, during an intense wildfire season in southern California that left sixteen people dead, over seven hundred buildings destroyed, and more than 500,000 acres burned. After the fact, it was established that California likely had enough resources to combat the separate wildfires; what made the season particularly catastrophic was a lack of communication among the local, state, and federal agencies, and the resulting inefficiencies.

As the California state firefighting program, FIRESCOPE, described it,

> At the time, the sky was full of giant smoke columns and fire apparatus were passing each other on their way to incidents, with some going north as others headed south. Individual Command Posts and fire camps were established by multiple agencies for the same incident.[5]

By 1973, southern California had developed the forerunner to the federal ICS that is still in place today, in order to create a "system which would provide uniform terminology, procedures, and incident organization structure required to ensure effective coordinated action when two or more agencies [in this case, all county and city fire departments, the state forest service, and the federal forest service] are involved in a combined effort."[6] This system has five principal divisions—Command, Operations, Planning, Logistics, and Finance—and is run by the Incident Commander (in the original context, the Fire Boss). During an oil spill, when several agencies share authority, the system is commonly known as the Unified Command.

Another aspect of the original plan was to develop technologies that would help improve the response to the crisis, including meteorology, data processing, and communications. In theory, the development of these technologies sounds like a place where science should come into play. But thus far, the integration of scientists into a crisis response has been far from seamless.

Why is this? When I asked Steve Lehmann this question, he responded:

> It is important to note that in an accident, it is not doctors who typically save your life. It is the EMT, who has broad rather than specific medical training and years of experience in similar events.
>
> A crisis has a beginning, middle and end. The beginning, in an oil spill and many other like events, is reactive, muscle memory. Detailed science is a luxury rarely afforded and rarely available. Immediate action, based on planning, experience, training, and creativity can significantly mini- mize the long-term impacts.
>
> The middle and end of a crisis is where detailed science is critical— cleanup methods and endpoints, rehabilitation and restoration, forensics, etc. For a scientist who is not regularly on the front lines, entering a spill is like joining a [game of] jump rope: timing is everything. The scientist doesn't control the rope but must coordinate with those who do and not interfere with the scientists who are already jumping.[7]

Steve went on to say,

> One issue that responders often have with academics is they don't want to share their data. They want something from us—say, a sample from the disaster site—but they don't want to give us anything in return. They don't trust us not to accidentally spill the beans on their next big paper. This can make cooperation difficult. You have to be able to trust one another.[8]

This doesn't mean there aren't independent scientists involved in a crisis response. After all, during *Deepwater Horizon*, NOAA reached out to me almost immediately, and they eventually created five new scientific teams to deal with the spill, such as the Flow Rate Technical Team and the Oil Budget Team. But there was a reason they reached out to some people and not others: the people who wound up on the inside were those who already had preexisting relationships. As Incident Commander Thad Allen was fond of saying: "You don't exchange business cards during a crisis."

Connections are important because they get you on board. But there's another aspect to this as well. Connections are important because, as Steve pointed out, you are more likely to understand and trust the people you are working with. You have already had a chance to develop mutual respect. A response is a team effort, and during *Deepwater Horizon*, the scientists who were on the outside, the ones who were frustrated that they had no way of communicating with the responders, in some cases seemed to be trying to sabotage the team.

The government missed an opportunity because it had no means of integrating new findings from the scientific community at large. But a segment of the scientific community also missed an opportunity because they interfered with the government response, made unfounded speculations that caused considerable anxiety among the affected public, and, most crucially of all, eroded the public's trust in the government to handle a crisis.

The Guerrilla Geochemist

When I received Steve Lehmann's first email, the one that said, "Lots of 'face time' on the news," I was flabbergasted. I thought I had been doing everything right. I thought I was doing the community a service. I had no funding to study the spill, and yet here was Steve, obviously annoyed that I was in the media and pushing me to share my data. We exchanged a few messages, and the mutual tension remained.

If we had never seen each other again, I think we both would have been happy. But the two of us worked in the same field, in the same part of the world, and bumping into each other was inevitable. In the years that followed, every time I heard his name come up in conversation, I would say to myself, "What a hack!" Finally, in 2005, my friend and colleague Captain Rick Gaines of the Coast Guard told me to stop badmouthing him. "Listen,

A crisis is not the time to exchange business cards.

Figure 3.1 In the heat of the moment, success favors those with preexisting relationships.

Chris," he said, "Lehmann is a good guy. When something goes wrong, he's the person you want on your side. You two really need to sit down and have a drink together."

And so, during an oil spill response meeting at the University of New Hampshire sometime later, we did. Steve ordered a beer and got straight to the point. "Let me explain why I'm so pissed at you," he said.

> You have no idea of the power you have to ruin my day. When a scientist comes out of nowhere and tells us we're doing it wrong, or makes some bizarre claim, it keeps us from doing our job. We have to answer to that. Our boss wants to know, "Is what he says true?" The local government officials want to know, "Is what he says true?" The media wants to know, "Is what he says true?" And every hour, every day that goes by when we aren't doing our job—because you're out there talking about naphthalene—that's another fifty miles of coastline that gets oiled. We are trying to stop that from happening, and you guys are out there trying to throw a wrench into the machine.

He took a long pull on his beer, then placed the glass back on the table.

> Listen, I can't tell you not to speak to the press. I know you're a talented guy. But if you want to share your insight, you're better off calling me on my cell. Try to be more mindful of what you say publicly. Because I don't want you to be a guerrilla geochemist. You can make a difference, but you need to do a better job.

We spent a long time talking that night. And thanks to that conversation, I no longer think of Steve as a hack—instead, he is a close friend. He may not have a research group or a PhD, but I know what his expertise is, and I've come to appreciate his values, know-how, and dedication. I may have published two hundred papers, but thanks to Steve's knowledge and experience, he has been able to save hundreds of miles of coastline. He has worked on every major oil spill in the past thirty years, from *Exxon Valdez* to the First Gulf War. What matters more?

How does this relate to networking? I would argue that the most important connections in my life as a scientist have all come out of local events, such as the *North Cape* spill in Rhode Island or the *Bouchard 120* spill in Buzzards Bay. Starting at a local level is how to gain the experience that you need to communicate effectively.

During *Deepwater Horizon*, I was one of the few scientists who had the cell number of someone on the response team. And one of the reasons I had an "in" was because Steve and I knew each other. I could have spent the rest of my life thinking that NOAA's Region 1 science support coordinator was a hack. But because a friend persuaded us to sit down together, we went on to establish mutual respect and become colleagues. We knew we would continue to cross paths in the future, so we learned to value the other's expertise. It was clearly more beneficial for us to work together rather than remain a thorn in the other's side. With networking, it's crucial to remember that, at first, you may not like the person on the other side of the table. And they may not like you. But if you have to collaborate at some point in the future, it will still be easier if you already know each other.

Today, I always tell junior scientists: find out who the local players are in your town and reach out to them. Meet with a reporter at your local NPR (National Public Radio) station. Find out more about what the most important part of her job is. Find out what she values. And then explain to her what you're excited about researching, and what your value system is. Once you figure out what sort of stories she covers, and once she figures out what sort of expertise you have to offer, she may come to think of you as a source, and down the road, she may go on to introduce you to other journalists. But first, you have to reach out to her. Not only do you need to understand the needs and goals of the different stakeholder groups, but you have to meet them in person.

The same goes with emergency responders, like the local fire chief or Coast Guard officers. Tell them, "Hey, I'm a local resource. I'm on a first-name basis with the atoms. What can we do for each other?" People might not want to listen to you give a lecture, but they will always be interested in trading information. During a crisis, both sides will gain credibility through association.[9]

Invite your connections to seminars or lab groups. More often than not, visitors to our world will tell us what they think we are doing wrong and what they wish we would do better. This is also an excellent opportunity for us to explain how science works (see the Appendix).

If the city council is discussing pesticide use, or some other topic that relates to your knowledge, go to the meeting, introduce yourself, and state your opinion. The local government may value what you have to say, and you will be showing your fellow citizens that you care about local issues. Maybe they'll disagree with you, but that's part of the democratic process. Go out and have coffee with one of the science teachers at a nearby high school and learn how you can get involved with the students. There are all sorts of low-risk opportunities at the local level. And each opportunity will teach you more about how to communicate, while at the same time building your network of trusted acquaintances. No one goes from graduate student to insider. This is a long process, and you have to be patient.

Notes

1 M. Blumer et al., "The West Falmouth Oil Spill," Woods Hole Oceanographic Institution, September 1, 1970, https://apps.dtic.mil/sti/pdfs/AD0713947.pdf; E. G. Ward, "A Silent Fall: The Story of the West Falmouth Oil Spill," *Woods Hole Museum* 22, no. 1: www.woodsholemuseum.org/oldpages/sprtsl/v22n1-SilentFall.pdf.
2 I am on good terms with several consultants; however, this was not someone I knew.
3 Lubchenco et al., "Science in Support of the Deepwater Horizon Response," *PNAS* 109, no. 50 (2012): 20212–21; https://doi.org/10.1073/pnas.1204729109.

4 S. Hemphill, "Researcher Crusades for Policies to Protect Water: Profile of Dr. Deborah Swackhamer," *Agate*, March 24, 2016. www.agatemag.com/2016/03/researcher-crusades-for-policies-to-protect-water-profile-of-dr-deborah-swackhamer.

5 "History of ICS," https://firescope.caloes.ca.gov/SiteCollectionDocuments/ICS%20History%20and%20Progression.pdf.

6 U.S. Forest Service, "FIRESCOPE, a Record of Significant Decisions," 1981.

7 Personal communication. Steve also noted: "By the same token, the interference of less experienced scientists can make matters worse. For example, in the early days of the Exxon-Valdez spill, dispersants were considered, but the use of such a technology was publicly fought by Dr. Riki Ott, a biologist-turned-fisherwoman. Having no background in oil spill response, she successfully delayed the tactic. Three days later a large storm distributed the oil throughout Prince William Sound only to return in ocean convergences. It was these convergences that had the most impact on rafting birds."

8 Personal communication.

9 Once again, thanks to Steve Lehmann for contributing his words and thoughts to this paragraph, in particular the idea of credibility through association.

4 Who Writes the Narrative?

Less than a week after the *Deepwater Horizon* explosion, the oil slick extended as far as the eye could see, and yet there was little evidence of a major cleanup. Why was the government moving so slowly? On April 25, BP made the startling announcement that its wellhead on the seafloor had ruptured, and oil was leaking into the Gulf at the rate of one thousand barrels per day. Together with the Coast Guard, the company declared their containment plan, which consisted of in situ burning, booms, skimming, and the application of dispersants, which would help break up the surface oil.

This was the Unified Command in action. Most of the public had no idea, but the headquarters in Robert, Louisiana, had been up and running since April 23, the day after the rig sank. Members included representatives from the following federal agencies: the Environmental Protection Agency (EPA); the Coast Guard; the Departments of Commerce, Agriculture, Defense, Homeland Security, Interior, and State; the National Oceanic and Atmospheric Administration (NOAA); the Fish and Wildlife Service; the National Park Service; the Minerals Management Service; the US Geological Survey; the Centers for Disease Control and Prevention; and the Occupational Safety and Health Administration. All of these people answered to Coast Guard commandant Thad Allen, who was named Incident Commander on April 30.[1]

Crucially, BP and Transocean (the offshore drilling contractor) also had a seat at the table. Among the various aspects of a crisis response that the public neither understood nor agreed with, the inclusion of BP as part of Unified Command was the most significant. To be clear, federal law states that the Responsible Party in an environmental disaster is in charge of their own cleanup: they run the Operations division. The government, meanwhile, oversees the Planning, Logistics, and Finance divisions. However, because the role of the Responsible Party was misunderstood, it was generally perceived that BP was not being held accountable. Indeed, because they were in charge of Operations, it often appeared as if they were calling the shots.

After just one week, the slick was within twenty miles of the Louisiana coastline, and moving rapidly toward the Mississippi Delta.

DOI: 10.4324/9781003341871-7

On April 28, NOAA stepped in and readjusted the estimated flow rate from the leaking wellhead: it was not one thousand barrels a day, it was five thousand barrels per day.

On April 29, Louisiana declared a state of emergency, and fisheries began to close.

And on April 30, just ten days after the rig exploded, the oil began to wash ashore in Venice, Louisiana.

One of the most confounding problems with the *Deepwater Horizon* response was the public's lack of trust in both the government and BP to do the right thing. This mistrust was baked in from the beginning, in large part because no one understood the established role of the Responsible Party. After all, BP's well-documented safety lapses had resulted in the deaths of eleven local workers. How could they be expected to do any better when it came to stopping the spill? And how could the government be trusted after the failures of the federal response during Hurricane Katrina, whose wounds were only just beginning to heal in coastal communities?

Thus, not only was the affected public skeptical of those charged with solving the problem, the concerned public was also skeptical. And that included scientists. And some of these scientists—those with expertise who were not called on to assist with the response—considered it to be their duty to play the role of sheriff.

I am not going to criticize those who challenged the official narrative, because once upon a time, that was me. From our perspective, we were just being rigorous—that's what we are trained to do. It wasn't until I sat down with Steve Lehmann that I learned to see the view from the other side.

In early May, when everything was still going sideways, a lot of journalists were looking for more information from oil spill experts. The paper I had written in 2002, about the residual compounds that had persisted for decades after the *Florida* spill in West Falmouth, began to resurface. And once reporters got wind of the findings, they had their angle: *Deepwater Horizon* oil was going to last in the Louisiana salt marshes for another forty years.

Soon after this, a reporter from a major newspaper visited and asked me to take her on a tour of the salt marshes where the *Florida* spill had occurred. I reminded her during the tour,

> This particular spill and outcome is an anomaly. The oil here that did not decompose was one-tenth of one-tenth of a percent. Where we find oil now is a very small area compared to the area initially oiled. This is just one case study. There are other diesel fuel spills where you can't find any residue at all after two years.

Of course, when the story came out the next day, it was all about how the Gulf of Mexico is going to be oiled for the next forty years. Did the reporter forget to include my clarification? Did a subeditor cut it out of the story during layout in order to meet the word count?

Later that month another reporter contacted me for an interview, and we went through the same process. I explained the residual effects of the *Florida* spill and clarified that this was an anomaly. In our email exchange I wrote: "I beg of you that if you include any story on this spill that you mention that other oil spills have acted on salt marshes and been much less detrimental ... underscoring that all oil spills are not the same."

When the article came out a few days later, the headline included the phrase, "ecological damage could last decades." I was quoted in the article, but my plea for balance was ignored. Instead, a much catchier line that I had used to describe the impacted zone was featured: "The marsh is still waging chemical warfare several inches below the surface."

I understand why the reporter went with that quote—it supported his angle and, frankly, it sounded good. But by now I had begun to realize that few people in the media were going to run a story that included a best-case scenario or, at the very least, a range of different possibilities. It was a classic example of bad-news bias.[2]

So by the time another journalist I had been in touch with wanted to visit, I had already decided to opt out. A different scientist gave the tour instead. I was mindful that there was an impact to these stories that was not helping the response; though, in this case, there was little I could do.

Box 4.1 Staying Coolheaded in a Crisis: What We Can Learn from Spock

Back in the days when there were only three channels on TV, my brothers and I had a favorite after-school routine. Every day at 4 o'clock, my older brother Joe would climb up on the roof and fiddle with the antenna until the reception improved enough so that we could watch our favorite show: *Star Trek*.

I've watched all of the original *Star Trek* episodes hundreds of times over the years, and I still enjoy them. Why? Part of the reason is because the basic plot is almost always the same: Captain Kirk and his crew fly around on the USS *Enterprise* and try to solve some sort of crisis. And here's the thing: we can learn a lot about crises from *Star Trek*. Specifically, we can learn a lot from the *Enterprise*'s star communicator: Spock. Everybody remembers Spock—green skin, pointy ears, half-Vulcan. He dropped lines like, "Emotions are alien to me, I'm a scientist."

In the episode "That Which Survives," the *Enterprise* and its crew are mysteriously teleported one thousand light years away from their original location. Spock, always a stickler for the details, jumps in to the discussion to clarify that they haven't moved one thousand light years, but 990.7. After Mr. Scott disputes the claim as utter nonsense, Spock again jumps in to remind him that he is the one who is being completely illogical, since the ship did indeed move to a new location.

If you watch this scene for yourself, you'll see that Spock does the stereotypical, annoying scientist thing. He's cold. He's rude. He corrects someone. They were one thousand light years away from where they should have been, and he said 990.7? Who cares?

Well, while most viewers might mock him for saying 990.7, that number actually means something. Spock is not flashy, and he's definitely not sexy. But he gets the job done. Even as we sense panic rising in the other crew members, Spock remains cool and unwavering. He stays on message: he reminds the others—not once, but twice—that they are exactly 990.7 light years away from where they should be.

These are all important traits in a crisis. Could he have been a little nicer? A little more like Dr. Fauci? Without a doubt. Spock is certainly not the person you would want to run a press conference. But he is exactly the person you would want to inform the person running the press conference.

What's important in a crisis is that you deal with the facts: this is what we know, this is what we don't know. This is what's changing. This is what's up for debate. Spock doesn't display any bias, and he never tries to please anyone either. That, too, is important. Too often, scientists who are in a crisis mode and aren't well prepared try to make other people happy. They say what they think the other person wants to hear instead of staying on message.

Despite Spock's claim that emotions are foreign to him, the truth is, he is an unbelievably passionate scientist. The key is, his delivery is dispassionate. And that is what makes him such a powerful communicator when everyone else is starting to lose their cool.

The Dangers of Noninformation

I am not taking an antimedia stance here, but it's important to remember that it is okay to say no to an interview. Situational awareness is a big part of effective communication. What will the consequences of your statements be? If you appear in a news story, you will be cited as an expert, and people will likely take your words seriously. In this particular case, I didn't want the media to continue to misrepresent my findings, because I felt that people who lived in the Gulf states would get the wrong idea.

While *Deepwater Horizon* was an unusual event—a once-in-a-lifetime disaster—it is an excellent example of science communication because it accentuates all of these nuances. In another situation, you might not see the harm of speculating out loud. But in this case, we can see how each mistake by the various stakeholders went on to play a huge role in the months that followed. Various misstatements led to tensions—between independent scientists, responders, the media, industry, and the public—that were long-lasting and detrimental.

Communication in a crisis is tricky. BP made the first mistake of all, when they insisted that only one thousand barrels of oil per day was leaking from the wellhead. Even after NOAA upped the estimate to five thousand barrels per day, it seemed clear to many people from the images on TV that *a lot* of oil was flowing out of the ruptured well.[3] Way more than five thousand barrels per day (the final figure was eventually estimated to be roughly a dozen times the initial government figure[4]).

So why wasn't Unified Command trying to figure out the exact rate of flow instead of lowballing the estimate? Because at this point they didn't need to know the exact number. They had already pulled every alarm in the nation. Their primary concern was plugging up the well and containing the oil that had already spilled. The flow rate calculation was just one priority among countless others.

But the public saw it differently. The low estimate cemented in people's minds that neither BP nor the government was going to tell the truth. And when access to the wellhead was closed off to anyone not involved in the response, this compounded the public mistrust. Why was the government closing off access to the wellhead? It had to be because they wanted to keep independent scientists from determining the actual flow rate, right?

NOAA director Jane Lubchenco later revealed the actual reasoning behind this decision:

> Academic and independent researchers wanted access to the well site at depth, but their presence had strong potential to interfere with operations to stop the flow of oil. Remotely operated vehicles (ROVs) controlled from the surface with acoustic commands were attempting delicate maneuvers at depth in the dark. Scientific ROVs had strong potential to interfere physically or acoustically with response ROVs. After one response ROV accidentally bumped into and dislodged the riser insertion tube tool (an early device for collecting oil inside the riser), the NIC declared a "no go" zone in the critical area around the well. Permission to enter that zone was then allowed by the NIC only if activities would not interfere with response operations. Although some researchers understood and respected the "no go" zone, others complained that a heavy-handed government was preventing science from proceeding.[5]

Misunderstandings such as this one led to increasing conflict between independent scientists and the government, and a growing appetite among the public for news stories that challenged the official narrative. This should have been the moment for scientists to step in and say, "Hey, these are the facts. This is what we know, this is what we don't know. I recommend this, and I would be happy to expand."

But instead, certain scientists joined the fray and started to spread noninformation—that is, worst-case-scenario hypotheses that had no supporting evidence. The hurricane season was one popular topic that was picked up in the media. Some experts claimed that high winds were going to blow the oil everywhere, making the coast uninhabitable for years to come and even ruining crops inland. Such statements amplified the pervading sense of fear and catastrophe among an affected public who had only just recovered from Hurricane Katrina.

In *A Sea in Flames*, the ecologist Carl Safina included numerous examples of those who were contributing end-of-the-world noninformation to media stories during *Deepwater Horizon*, such as the following from a Texas A&M professor:

> The threat to the deep-sea habitat is already a done deal, it is happening now. ... If the oil settles on the bottom, it will kill the smaller organisms, like copepods and small worms. When we lose the forage, then you have an impact on the larger fish.

Safina went on to comment:

> Yeah, maybe. But, "a done deal"? Really? How much oil would have to settle? How densely? ... I am deeply distressed about the potential damage to wildlife and habitats—but I find myself uncomfortable with all the catastrophizing. A "done deal"—that's not very scientific. Especially for a scientist. Many scientists—and as a scientist it hurts to say this—are being a little shrill.[6]

Reading through the original Associated Press article, which first appeared on May 5, 2010, and was picked up by at least a dozen newspapers and websites in the days that followed, one gets the sense that the entire Gulf was going to be contaminated, and then, by way of the Loop Current, the oil would make its way around the tip of Florida, polluting beaches and killing marine life along the entire East Coast in the months that followed. In hindsight, we know this didn't happen. Could it have happened? Perhaps.

So, when do you draw the line between a hypothesis and a definite outcome? Is there value in stating that the Loop Current *might* cause the oil to flow up the East Coast? There is, but perhaps not publicly. Information on currents, the marine food web, and the location of deep-sea coral is all of interest to responders. Here, we see how a lack of a clear communication channel between the government and independent scientists not only frustrated both sides, but also had a negative outcome for the public.

The scientists quoted in this article thought they were making an important announcement—and they were—except that they were making it to the wrong stakeholder group. They didn't realize how these unfounded statements would affect all the fishers whose livelihood was based in the Gulf. How all the tourism-based businesses in the Gulf states—and even along the East Coast—would be impacted by such dire predictions. And, it bears repeating, these were predictions that did not come true.

Box 4.2 Cocodrie

In the early weeks of *Deepwater Horizon*, my role as a government consultant had me chained to a desk in Cape Cod, and it wasn't until mid-May that I was finally able to take the first of many trips to the Gulf. I flew into New Orleans, rented a car, and drove down to Cocodrie, Louisiana, a small fishing and shrimping town at the edge of the marshlands. There I met with Nancy Rabalais at LUMCON (Louisiana Universities Marine Consortium). Afterward, I asked if anyone could take me out to collect some samples, and a local who also worked for LUMCON gave me a ride out to see the oil in a twenty-foot powerboat. This guy had lived in Cocodrie his entire life. Even though New Orleans is only eighty-six miles away, he had been there just once, to see a Saints football game.

In order to get out to the open water, we had to navigate our way through a maze of creeks and marsh highways, and as we were chugging along he told me all about the coastline. The things he knew about geomorphology were astonishing. *This area is only now coming back after Hurricane Gustav*, he would say. *And this area over here has been moving every day.*

It was a great tour, but when we made it to the open water, that's when we saw the oil. And at that moment he choked up, because his whole life—his whole life—was being painted black right in front of him, and there wasn't a thing he could do about it. Not only could you see the oil, but the smell permeated the air. It was everywhere.

After he finished taking me around, we went back to Cocodrie, said goodbye, and then I walked back to my rental car. And I remember just sitting there for the longest time. I was trying to get over the fact that I wasn't a scientist anymore. I wasn't just collecting samples and analyzing them and writing papers. I was in the middle of this guy's life.

And the thing was, he was really appreciative that I went down there. I wound up going back to Cocodrie several times during the spill. At first, I was worried that I would be viewed as a Yankee or an outsider. But actually, the affected public just wants to know that people care about them. One time, a local police officer even offered me a sandwich that his wife had made, simply because he was so grateful that I was open to discussing my work with him. I wasn't even doing anything—just collecting samples—but they needed to know that there were people out there who were trying to help. That their home wasn't going to be covered in oil for another forty years.

It's one thing to think about how your words might impact someone. But if that person is a stranger, you're probably only going to think about this in an abstract way. It's another to actually talk to the people who are impacted, to see how their lives have been changed, and listen to what they have to say. The better you know the affected community, the more on target your communication will be.

The Dispersants Debate

Another popular topic that gained traction at this time concerned the use of dispersants. Dispersants are a part of the response toolbox for oil spills, because they help break up oil slicks into smaller, less buoyant droplets, which result in less damage to coastlines and flora and fauna that lives at the surface.[7] They are most often compared to the effect dish soap has on a frying pan coated with bacon grease.

Like fire retardants that are sprayed from planes during a wildfire, dispersants are applied to the sea's surface to slow an oil slick's spread. They are not a perfect solution, because in breaking up the oil and sending it beneath the surface, responders are only transferring the pollutant to a new area rather than removing it. However, the volume of water in the ocean is such that the oil is dispersed in much smaller concentrations, and the smaller droplets that the dispersants create are much more appealing to the microbes that eat petroleum.

But almost immediately, concerns were raised: Are dispersants toxic? And if so, when they are sprayed from planes, could they affect the thousands of local fishers who were now employed placing booms out at sea? If they were sprayed close to the coastline, could they blow inland and affect the people living nearby?

This was obviously not an ideal situation. When many people learned about the dispersants, they thought: *the government is taking another toxic chemical and adding it to the first one. Their actions are making a bad situation worse.* This was, in fact, the same sort of emotional response that drove vaccine hesitancy during the COVID-19 pandemic. Even though dispersants were already a known, proven quantity that had been successfully used to control other oil spills, and even though the ratio of dispersants to oil was approximately 1 to 20, the affected public quickly latched on to the idea that the application of dispersants was even worse than the oil itself.[8]

The truth is, there are no perfect solutions in an environmental disaster. Responders are trying to obtain a net positive in as short a time as possible, and in doing so, they have to use all the tools available to them. Firefighters use flame retardant because they are trying to save a specific area that has been deemed important—such as a residential community. This doesn't mean that they aren't cognizant of the effect of flame retardants on riparian zones. It just means that, at that particular moment, using flame retardants was the best of several bad options.

During *Deepwater Horizon*, the amount of oil that was spilling into the ocean each day was so massive that responders had to use all the tools that were available to them. No single technique—skimming, booms, fires, or dispersants—was going to work on its own. Eventually, the government even decided to apply dispersants underwater in order to prevent the more volatile components of the oil from reaching the surface.[9] But even this action was misconstrued—while the goal was both to improve air quality and to keep the oil from washing up along the coasts and into the marsh, where it would likely do the most damage, some people suspected that, once again, the government and BP were trying to hide the extent of the problem.

To be sure, no one knew what the outcome of discharging 1.6 million gallons of chemicals into the ocean to disperse the oil was going to be. There would certainly be effects on the ecosystem. But to what extent? And how long would they last? That was an unknown. The problem is that, during a crisis, we don't have the luxury to sit around and collect samples to learn more about the outcome. Our level of certainty that a particular course of action is the right one is not 99%. It may only be 70%. But in the moment, 70% is as good as we can get.[10]

By this point, the fault lines that had formed between the government response and independent scientists were too far apart to be reconciled. As I had discovered with Steve Lehmann during the *Bouchard 120* spill, responders are not happy when independent scientists criticize their efforts to put out a fire, especially while the fire is still raging. The growing media uproar over dispersants and possible toxicity increasingly interfered with the Unified Command's ability to focus on stopping the leak and containing the spill.

And then, suddenly, the dispersant question evolved into something else. On May 15, a team of oceanographers on board the *Pelican*, who were already out conducting research in the Gulf, made another public announcement: Hey, some of this dispersed oil isn't making it up to the surface. It seems to be forming giant subsurface plumes that are miles long.

Subsurface plumes of oil? Holy heck! What's that? Were massive rivers of crude as thick as Hershey's syrup running through the Gulf? If tensions between responders and independent scientists were already stretched thin, this particular announcement took them to a new level.

Notes

1 Thad Allen, "National Incident Commander's Report: MC252 Deepwater Horizon," US National Response Team, October 1, 2010, www.nrt.org/sites/2/files/DWH%20NIC.pdf.

2 As David Leonhardt wrote of the bad-news bias during the pandemic, "About 87% of Covid coverage in national U.S. media last year was negative. The share was 51% in international media, 53% in U.S. regional media and 64% in scientific journals. Notably, the coverage was negative in both U.S. media outlets with liberal audiences (like MSNBC) and those with conservative audiences (like Fox News). ... As Ranjan Sehgal, another co-author, told me, 'The media is painting a picture that is a little bit different from what the scientists are saying.'" See D. Leonhardt, "The Morning Newsletter: Bad-News Bias," *New York Times*, March 24, 2022.

3 To be sure, it would have been quite difficult to estimate the actual rate of flow just by looking at an image, because a lot of natural gas was mixed in with the oil. The gas was expanding quickly as the pressure decreased with the elevation change (the actual oil reservoir was roughly 2.5 miles, or 13,360 feet, beneath the sea floor), and this expansion created the impression that there was more volume than there actually was.

4 See M. McNutt et al., "Review of Flow Rate Estimates of the Deepwater Horizon Oil Spill," *PNAS* 109, no. 50 (2012): 20260-67, https://doi.org/10.1073/pnas.1112139108 for a lengthy discussion of the flow rate over time.

5 Lubchenco et al., "Science in Support of the *Deepwater Horizon* Response," *PNAS* 109, no. 50 (2012): 20212–21; https://doi.org/10.1073/pnas.1204729109.

6 C. Safina, *A Sea in Flames* (New York: Crown, 2011), 81.

7 See National Academies of Sciences, Engineering, and Medicine, *The Use of Dispersants in Marine Oil Spill Response* (Washington, DC: National Academies Press, 2020), https://doi.org/10.17226/25161.

8 The earliest dispersants were, in fact, quite toxic. However, even though the science and the product have improved over the years, public awareness never caught up. A lot of the bad rep associated with dispersants stemmed from the 1967 *Torrey Canyon* spill off the coast of southwest England. As ITOPF notes: "A distinguishing feature of the TORREY CANYON response operation was the excessive and indiscriminate use of early dispersants and solvent based cleaning agents, which caused considerable environmental damage. The dispersants were generally successful at their task of reducing the amount of oil arriving ashore and subsequently expediting onshore cleanup operations, but they were considerably more toxic than those used today and were applied in far greater concentrations, often being poured undiluted on slicks and beaches. Many of the detrimental impacts of the spill were later related to the high volume, high concentration and high toxicity of the dispersant and detergents used." See www.itopf.org/in-action/case-studies/torrey-canyon-united-kingdom-1967. See also Roger Prince's discussion of the use of dispersants over the past half century: "A Half Century of Oil Spill Dispersant Development, Deployment and Lingering Controversy," *International Biodeterioration & Biodegradation* 176 (2023): https://doi.org/10.1016/j.ibiod.2022.105510.

9 J. S. Arey and C. Reddy, "Did Dispersants Help Responders Breathe Easier?," *Oceanus*, August 28, 2017, www.whoi.edu/oceanus/feature/did-dispersants-help-responders-breathe-easier-at-deepwater-horizon.

10 In my opinion, the people who made the decision to use the dispersants were the right ones; they had the proper training, knowledge, and appreciation of the stakes, and were well versed on the long body of research on the topic, which includes two reports by the National Research Council; one in 1989 and the other in 2005. See: https://nap.nationalacademies.org/catalog/11283/oil-spill-dispersants-efficacy-and-effects and https://nap.nationalacademies.org/catalog/736/using-oil-spill-dispersants-on-the-sea.

5   Winner Takes All

The announcement of subsurface plumes of oil may not have pleased the response team, but it certainly caught the attention of the scientific community and the media. But just what is a subsurface plume?

We now know that when hot oil and natural gas erupt from the sea floor, the force of the initial interaction with the cold water breaks up the fluid and gas into smaller components. You may have oil blobs containing gas bubbles, or gas blobs containing oil droplets. As the plume rises, more and more water becomes incorporated into the plume, and the oil and gas continue to break down into smaller and smaller compounds. Eventually, the largest components rise to the surface, while the smallest ones reach neutral buoyancy and begin to drift in the underwater currents. Some compounds simply dissolve in the water column.

So, if you were to look at a sample taken from one of these plumes, it wouldn't resemble oil at all—for all intents and purposes it just looks like regular seawater. It was only in analyzing the composition that we could tell that it contained concentrations of compounds such as benzene, toluene, ethylbenzene, and total xylenes.

The discovery of the plumes was first made on board the *Pelican*, a research vessel that was preparing for an expedition into the Gulf right as the *Deepwater Horizon* rig exploded. The *Pelican* team, consisting of members of the National Institute for Undersea Science and Technology (NIUST), was originally slated to map the sea floor a mere nine miles from the *Deepwater Horizon* drilling site.[1]

After the news of the spill broke, however, the team shifted priorities: they decided to focus on "taking samples of sediment cores throughout the region, to develop a reliable baseline for future studies of oil that may settle to the sea floor."[2] While many stakeholder groups were frustrated with independent scientists throughout *Deepwater Horizon*, this is a perfect example of how academics are both nimble and able to contribute valuable information during an environmental crisis. Consider that the scientists on board the *Pelican* had been preparing for their cruise for at least a year in advance, and yet they were willing to drop everything and rerig their boat with no guarantees with regard to their original research plans—this was an incredibly selfless act.

DOI: 10.4324/9781003341871-8

Roughly two weeks later, on May 12, Arne Diercks and the rest of his team on the *Pelican* identified something that few people were expecting. A cluster of deep-sea sensors seemed to be indicating that there was oil flowing through the ocean depths.[3] Following further tests, the scientists hypothesized that this was a plume of hydrocarbons that had emanated from the wellhead and, instead of rising to the surface, was now drifting in the deep sea currents. Initial estimates put the plume at three miles wide, three hundred feet tall, and ten miles long. Were these plumes definitely emanating from the wellhead? If so, where were they going? And how would the environment and the marine food web react to them? These were the next big questions.

The news of the finding broke a few days later, on May 15, when the *New York Times* reported that, "there's a shocking amount of oil in the deep water."[4] When the *Pelican* docked the next day, its team found a crowd of journalists waiting for them on shore. From the very outset, tensions were brewing between the scientists and Incident Command. Vernon Asper, a professor of marine sciences on board the *Pelican* at the time, recalls that they were desperate for official guidance on how to discuss the plumes in public. As he later wrote, "If [the government] had simply included us in their overall plan, we would have been happy to support their perspectives, but instead, they gave the media the impression that they were covering up something. Which, perhaps, they were."[5]

The following day, the government issued a gag order to the *Pelican* team: the scientists were no longer to speak to the media of their discovery. The National Oceanic and Atmospheric Administration (NOAA) followed this up with a press release of their own, which stated that the *Pelican*'s findings were "misleading, premature, and in some cases, inaccurate."[6]

At the National Incident Command's (NIC) daily press conference on the same day, the response team—which, as we know, did not want to have their focus drawn away from stopping the leak—predictably wanted nothing to do with subsurface plumes. BP's spokesman, Andrew Gowers, simply replied, "We have no confirmation of that, but my observation as a layman is that oil is lighter than water and it tends to go up."[7]

Now, an industry rep offhandedly dismissing the findings of a team of scientists was probably not the best look for the company. But, with the benefit of hindsight, we can understand his perspective—like Steve Lehmann, the NIC did not want to have to deal with any guerrilla geochemist distractions. And, of course, Mr. Gowers's statement wasn't exactly incorrect—at that point in the disaster, most people did expect the oil to go up to the surface. But the company's stance was disingenuous. For weeks afterward, BP continued to maintain that "the plumes don't exist," even though there was mounting evidence that they did. This was yet another PR misstep that went on to severely undermine the company's credibility.

The first round of interviews given by the *Pelican*'s research team, meanwhile, had already gained traction with the press. In Dr. Asper's initial interview on NPR, he was asked about the possible threat of dead zones that might be developing because of the plumes. To this he responded, "We see low oxygen associated with the plumes. There's no question about that. Now, is this going to be critically low, low enough to impact the things that live down there? *That's something we just can't answer.*"[8]

Asper was being honest here, and that's good: he comes right out and says, "We don't actually know if this is a problem." However, there was a reason he had been asked this specific question. Following the *New York Times*' first article, stories had immediately broken claiming that dead zones—which would develop if the bacteria that break down the oil compounds used up most of the oxygen in the water, simply because the plumes were so large—were a done deal. And if these giant dead zones did develop, then untold numbers of fish and shrimp would die. This was an entirely reasonable hypothesis: dead zones form in the Gulf every year as a result of nitrogen runoff in the Mississippi River. However, as Asper stated, it was impossible to know whether or not the plumes from *Deepwater Horizon* would definitively lead to the same outcome. This uncertainty, however, did not stop the media from speculating that the Gulf fisheries were on the verge of collapse.

The Associated Press had already run a wire story on the plumes, quoting Samantha Joye from the University of Georgia, who was working with the *Pelican* team on land, as stating: "Oxygen levels in some areas have dropped 30 percent, and should continue to drop. ... It could take years, possibly decades, for the system to recover from an infusion of this quantity of oil and gas."[9] Again, this was an entirely reasonable hypothesis. However, like the Loop Current that threatened to take the spill around Florida and up the East Coast, we now know that it didn't happen. This was simply another example of the media's tendency to focus on worst-case-scenario speculation, rather than presenting the full range of actual possibilities.

The *Guardian* too ran a story on the plumes, with the following subhead: "Scientists believe marine 'dead zones' being created."[10] Somehow, the editor who wrote the headline for this article conveniently overlooked that no such statement was made in the article itself.

While scientists made a few dubious claims during the month of May, I don't think we are to blame for the inaccuracies in this particular case. Journalists saw a catchy angle in the plume story and inflated it, despite the clear lack of evidence supporting their conclusions. Responders, meanwhile, again missed an opportunity. Because there was no established channel between the NIC and independent scientists, the information on the plumes—which was certainly relevant to the spill—was not communicated in a unified, coherent way. The responders thus missed their chance to get out in front of the plume narrative.

But that's not all that was happening here. In the end, another story that was of equal significance to scientists wound up buried near the end of Dr. Asper's NPR interview. When the interviewer asked him what the next steps in studying the plumes were, Asper replied,

The first thing we're going to do is analyze our data and analyze the samples. And, of course, we're planning our next cruises. We're already making inquiries into finding ship time. It turns out that the limiting factor for studying this plume is the availability of research vessels. The research fleet in the United States for academic purposes has been dwindling over the last few decades, and *there just aren't ships available.* So we're having a hard time getting access to vessels that can take us out there.[11]

Any practicing scientist will recognize this situation: scientifically interesting topic, limited resources. And how do you think this played out?

Hunting for Plumes

The moment I heard about the plumes, I knew I wanted to study them. Part of the reason for this was because, just nine months prior, I had been off the coast of southern California with a team of scientists studying a similar sort of phenomenon: natural petroleum seeps in the ocean floor. They weren't coming out of the ground at sixty thousand barrels per day—the rate was more like half a barrel per day—but despite the reduced scale, there was overlap. We're still talking about hot oil that is being released into the cold, dark ocean depths.

In order to study these natural seeps, our team used two special pieces of technology. The first was a small mass spectrometer, which was about the size of a scuba diver's oxygen tank and which we used to sniff out oil compounds in the water and follow it back to its source. The second was an autonomous four-propeller robot called *Sentry*.

Basically, we would tell *Sentry* where to go, and it carried the mass spectrometer around until it found traces of petroleum compounds, which we then tried to follow back to the source. Our team, which consisted of Rich Camilli and Dana Yoerger from WHOI (Woods Hole Oceanographic Institution), Dave Valentine from University of California Santa Barbara, and John Kessler from Texas A&M University, was perfectly suited for doing research on the *Deepwater Horizon* plumes precisely because of the unique technology and methodology that we had developed during this cruise.

Even though the National Science Foundation (NSF) rarely funds oil spill research, this particular crisis was so extreme that they not only changed tack, but also allowed scientists to submit RAPID proposals, which are designed to cut through the red tape. Credit has to be given to the NSF, because this was an amazing adaptation on their part. Usually when you submit a proposal, it can take anywhere from six months to over a year to get the funding approved and have your research trip organized. But thanks to the RAPID grants, approvals were happening in a matter of days. The year after the spill, I was speaking at a conference and lamenting the fact that as a whole, scientists, including myself, weren't more coordinated. After the presentation, a European colleague came up to me and said:

> You know, you're being too hard on yourself. Given the circumstances, everyone did an amazing job. If this type of spill happened in Europe, we'd *still* be organizing our research cruise today. We never would have been able to get results so quickly, or get a boat out to sea in a matter of weeks.

Basically, we had all the right people who had previous experience working in deep sea environments. We had all the right technology and know-how that no one else had. The NSF had suddenly turned on the faucets to fund the research. But there was still a disconnect between independent scientists and the NIC. No one in the response community told us what would have been helpful. They could have given scientists a punch list that was tied to the NSF proposals: We need to determine how much oil is evaporating from the surface. We need to know how much oil is settling to the ocean floor. And so on.

Over the course of researching this chapter, I've spoken with many of the scientists who were out in the Gulf during the months of May and June. Most people agreed: amid the chaos accompanying one of the largest environmental catastrophes in the history of the United States, we still managed to deliver groundbreaking research. But there were multiple downsides that came with the chaos. We had trouble coordinating with one another.

We had trouble coordinating with the response team. Many of us didn't even know about the NRDA (Natural Resource Damage Assessment), whose mission was to assess the environmental impacts of the spill, and who could have used our help. Even today, many of my colleagues clearly recall the frustration and bitterness they felt—sometimes toward one another—during the spill.

A lack of coordination was inevitable. It is simply impossible to write a research proposal and prepare for a cruise in a matter of days. How could we have ever hoped to standardize our measurements, or ensure that we were covering all the bases for multiple months instead of just the first two? No one had the bandwidth to coordinate in depth with the other researchers.

But I would be lying if I said there wasn't another factor at play that further complicated our task. And it was this: competition. Because this is how the scientific model is structured. It's winner takes all—the people who make the big breakthrough, who publish the first paper on a new finding, get all the credit. And I, for one, absolutely wanted to be first.

If I wasn't first, I would have considered myself a bust. I was so well prepared for what was happening, and suddenly I felt like there were so many other people crowding into my field, that I had to show them all who was the expert.

We received our funding approval in days, which was certainly a record for me. Camilli, Yoerger, Sean Sylva (also from WHOI), and I formed one team. Dave Valentine and John Kessler each submitted their own proposals and formed another team. By my count, there were at least another five research vessels out in the Gulf in May and June. Some of us were funded by the NSF, others were funded by NOAA. All of us were interested in the plumes—albeit with entirely different research objectives.

The WHOI team was ready to go on June 15. I distinctly remember this day, because that morning I had to testify before a subcommittee in Congress on "Ocean Science and Data Limits in a Time of Crisis: Do NOAA and the Fish and Wildlife Service (FWS) have the Resources to Respond?" It was my second testimony that week. After the testimony was over, I changed into my work clothes, mailed my suit back home, hopped on a plane to Tampa, and by evening I was asleep on our boat. That's how fast everything was happening.

The first step in our plume-hunting expedition was to find how many plumes had formed and which way they were headed. The way we did this was by using a process known as "tow-yo." Tow-yoing involves towing different sensors worth several hundred thousand dollars behind a boat at a yo-yoing depth of several thousand feet. The tow-yoing was complicated by the fact that the well area was a complete madhouse—there were nearly one hundred vessels skimming and burning oil and gas in situ, siphoning oil up from the well itself, and working on a solution to cap the well. So the captain had to navigate around all of these ships on the surface, and the remotely operated vehicles (ROVs) beneath the surface, as he circled the well site, towing a steel cable that was over half a mile long from the stern of his boat. Soon after this, however, a new assignment came up. We had long known how important it was to establish a baseline analysis of what precisely was coming out of the well. Not only was it valuable for our studies of the plume, but it was also crucial for determining the amount of oil that was being spilled into the Gulf. While it wasn't an immediate priority, in the long run the total barrels spilled was a particularly important number, because it was directly tied to how much BP would pay in fines.

Prior to our research cruise, Rich Camilli and another WHOI engineer, Andy Bowen, had already been working with the Coast Guard on measuring the flow rate of the oil and gas coming out of the well using an acoustic measuring device. However, while they had managed to determine the overall volume, they still didn't know the exact oil-to-gas ratio. As it happened, another WHOI scientist, Jeff Seewald, had already developed the perfect tool for taking a deep-sea sample that could be used to determine this ratio. Thus far, however, the National Incident Command had denied us direct access to the well, afraid that we would interfere with their dual priorities of collecting oil and capping the leak.[12]

On the night of July 20, however, that all changed. The Coast Guard suddenly announced they wanted the WHOI team to collect a sample. They ferried four of us over to one of their response vessels—a three-hundred-foot boat called the *Ocean Intervention III*—where they snapped two lines on each person and winched us up six stories and onto the deck. In addition to Rich Camilli and myself, the team included WHOI geochemist Sean Sylva and Jarrett Parker, a Coast Guard officer who was also on board the *Endeavor*.

At that time, some of the oil and gas was being captured directly at the wellhead and brought up to the surface.[13] Because it was hard to sequester the gas, nearby boats were burning off the excess around the clock. The noise was deafening, and it was so hot we could actually feel the heat of the flames on our skin. As Sean Sylva, another geochemist who was with us, said: it was a "surreal environment: [there were] ships all around you, flames all around you, and bright lights and helicopters" were everywhere.[14] And it was in this context that they told us: "Okay, you have twelve hours in between operations to collect a sample. After that, the window is closed."

We had two isobaric gas-tight samplers (IGT) with us, which Jeff Seewald had originally designed to collect the fluids and gases that are naturally emitted from hydrothermal vents on the sea floor. The genius of the IGT is that it can collect material at high pressure and then maintain that original pressure until you get it back to the lab. In order to get the sampler down to the ocean floor, the *Ocean Intervention III* had an ROV hold it with a mechanical claw, and then, as a safety measure, they tied one end of a length of rope around the IGT and the other end to the ROV. So we had this incredibly sophisticated $40,000 device, with a titanium holding tank, that was being lowered five thousand feet to the ocean floor and held in place by nothing more than a robotic claw and a piece of rope!

After just five minutes, Sean looked like he had already lost five years of his life. And sure enough, halfway down on the first attempt, the ROV lost power and the claw dropped the sampler. Because the cameras also went out, no one knew if the sampler was still attached to the ROV or if it was now lying on the bottom of the ocean floor, lost for eternity. Thankfully, when we got the ROV back to the surface, the sampler was still there, dangling from the cord.

The second attempt was successful. We got the sampler down to the well, the ROV's lights switched on, and suddenly it was like we were right there—we could see this massive blowout preventer on a big screen, and there was the robotic arm, positioning the device into the middle of a gushing jet of oil and gas. The end of the probe was equipped with a temperature sensor, so we knew that when we found the highest temperature—which turned out to be 221°F—that we were in the center of the well. Sean slowly sipped a sample from the high-pressure stream over a ninety-second period, and we crossed our fingers that we were getting pure oil and gas without any seawater.

By the time we got the sampler back on board, it was the next morning, and our twelve hours were nearly up. Jarrett then transported the sampler back to land, where he met with another Coast Guard lieutenant. The two of them escorted the sample back to the WHOI lab, where it was stored in a padlocked refrigerator. Meanwhile, Rich, Sean, and I returned to the *Endeavor* and our original plume-mapping mission.[15]

Box 5.1 Lost in the Fog of War

During those two weeks at sea, if you had asked me on any day how I was doing, I probably would have said, "Awesome!" But the truth was, I was sleep deprived and jacked up on adrenaline the entire time. So much was going on, from the apocalyptic setting of a 3 a.m. sampling mission for the Coast Guard to the excitement of negotiating with *Science* via satellite phone, that I wound up making all sorts of mistakes. At one point, I was so tired and hungry that I even choked on fried rice while eating lunch. Sean Sylva had to use the Heimlich maneuver to save my life!

But in general, I had no grasp of the big picture. The entire time we were out on the *Endeavor,* we were traveling through oil slicks. Whenever we wanted to take a deep-sea sample, we would throw jugs of Palmolive on the ocean's surface and then spray it with a firehose in order to create a clean area that we could lower our sampling equipment through. This helped us avoid any cross-contamination with surface oil. And yet, despite sitting in slicks for two weeks straight, I never even thought to do something as basic as take a sample of the oil that was floating on the surface.

This was particularly significant, because two years later our team at WHOI stumbled upon a particularly important discovery regarding photo-chemistry, evaporation, and surface oil degradation. In retrospect, this particular finding turned out, in my mind, to be one of the three big outcomes from all of the *Deepwater Horizon* research. But when we wanted to do further investigations in the lab, we realized we had almost no samples to work with. I simply could not believe I had forgotten to do such a basic thing as take surface samples while we were out on our research cruise.

When I mentioned this to Sean Sylva, he replied: "Oh, I know where there's another sample. The ship's bosun [the deck boss] collected one because his wife just had a baby, and he took a sample as a souvenir, because he wanted to document to his kid where he was and what he was doing at the time of his birth."

So we tracked down the bosun and drove to his house in Rhode Island. He was nice enough to share some of his sample with us, and we wound up naming it Abe, in honor of his son. Because I was in the fog of war, and not thinking clearly, this was the only sample of surface oil that I managed to obtain from that particular two-week research cruise. I wasn't trained to have a funding proposal approved in a single day and to head out for research a week later. I wasn't trained to publish our findings less than two months after that. It's just not the timescale that science operates in.

So now I always tell people, when you're in the midst of a crisis, make sure you write everything down. Check in with someone on a daily basis. Talk to them about what you're thinking about, what you're doing. More importantly, tell them *how* you're doing, both mentally and emotionally. Ask them what they would be doing if they were in your shoes. Because no one did that for me, and it proved to be a mistake.

The plume research went even better than we hoped. Using the techniques we had developed in California and the information we had gathered while tow-yoing, we were essentially able to throw *Sentry,* our underwater robot, and the mass spectrometer into the ocean where we knew the hydrocarbons would be. *Sentry* then mapped the plume by itself. If the mass spectrometer registered that it had moved out of the plume, it would tell *Sentry* to turn back and head in the other direction. The whole process of zigzagging across the entire plume took just three days—if we had done it in the conventional way, gathering this much data would have taken months.

Our final measurements were a plume that was one mile wide, six hundred feet high, and at least twenty-two miles long. Along the way we had collected a bunch of samples from different parts of the plume so that we could analyze them back in the lab.[16]

By this point, we knew that our findings were white hot. In fact, we thought that our paper was going to be so good that I called the editor of *Science* from the boat via satellite phone and wound up negotiating with him to get it fast tracked for the next issue. But even if the plumes were generating plenty of interest in the scientific community, the responders *still* weren't impressed.

Looking back, I realize this was a low point for me. I was in full-on scientist mode. I was no longer thinking about the people who lived along the coast and how their lives had been ruined, or about the priorities of stopping the leaking well and containing the oil that couldn't be captured. I was only thinking about our results, and being the first to publish a paper.

A decade later, I was speaking with a friend and he told me that in the midst of everything that was happening, I failed to acknowledge him in our first publication. He still felt a twinge of bitterness years later; the fact that there were so many crossed wires of this kind underlies just how quickly and chaotically things were unfolding at the time.

Another friend, Dr. Deborah French McCay, an established oceanographer who was working as an environmental damage consultant for the NRDA (Natural Resource Damage Assessment), expressed frustration with both the responders *and* academics. As a fellow scientist, I found her comments particularly eye-opening.

Figure 5.1 A crisis is never scheduled, often chaotic, and provides a breeding ground for actions that divert attention to the most pressing issues.

One of the first things she said to me was this: "Right when we started to figure out what sort of work needed to be done, you guys went and took all the good boats. And you were all studying the same thing."

Just like Dr. Asper had mentioned in his May interview with NPR, everyone felt the pressing need to get out into the Gulf immediately, and there simply weren't enough research vessels to go around. Our particular boat, the *Endeavor*, had sailed down all the way from Rhode Island. In the race to get funding and organize research trips, all we thought about was being the first. The first to get a boat. The first to make a breakthrough. The first to publish a paper.

This may be an oversimplification—I know my colleagues feel quite strongly that we were not "all studying the same thing." But it's important to remember: crisis events will hamper our ability to coordinate. And Debbie's larger point was this: the assessment of environmental impacts during a crisis is a crucial task, and she was unable to get out to sea reliably and take as many samples as she needed. Meanwhile, throughout May and June, there were numerous other boats that were all in the same location, and that were all studying some aspect of the behavior, fate, and effects of hydrocarbons that did not surface.

Were scientists indifferent to the needs of the NRDA? Certainly not. Many of us were in contact with NOAA at the time. The WHOI vessel had one NRDA researcher on board. And Dave Valentine told me,

> One of my biggest regrets was that we didn't get NRDA support on our cruise. They were going to send someone and then we sailed with an empty bunk. ... [At the time,] I didn't know what NRDA stood for, and my cell phone didn't get reliable service in Mississippi. Both hindered my ability to coordinate.[17]

Figure 5.2 Knowing who to contact—and how—is a key requirement for success when communicating science.

Clearly, interagency communication and coordination were an issue. While the NRDA did eventually get access to a boat (their first vessel, the *Pisces*, was immediately requisitioned by the responders for acoustic monitoring), it turned out not to be a research vessel at all, but instead a work boat that had to be MacGyvered at the last minute into something that at least had the modicum of necessary equipment on board. Eventually, it too was requisitioned by BP.

By the time Debbie was able to get a research vessel that was going reliably out to sea, the well had been capped and—from an academic's perspective—it was too late to bring any of us on board as chief scientists: many academics were already back in our labs, analyzing results and writing papers. As she wrote to me in a later email:

> My beef with some academics: they were so busy trying to make a name for themselves and get their opinions published that cooperation with the NRDA was secondary and not of interest. They especially did not like having to remain confidential on findings.[18]

I am not going to speak for my colleagues—Dave Valentine and John Kessler, for instance, did go on to assist the NRDA with chemistry—but I know that this criticism, in my case, was certainly true. I *was* trying to make a name for myself. I *was* absolutely zeroed in on my own research. I also went on to write seven op-eds over the course of the spill. In short, by the time the well was capped, I was fully aware that I was going to have a bigger impact back at my lab than out at sea. I don't remember if the NRDA contacted me at the time, but I am fairly confident that, if they did, I would have said, "Sorry, I'm not available right now."

Just like business and sports, competition is an integral part of science. The idea that competition drives innovation and hard work is deeply ingrained in the scientific ethos. Tesla and Edison, Newton and Leibniz, Pasteur and Koch, Darwin and Russell: the list of competitors—and their achievements— goes on and on. Competition also provides another service: it strengthens conclusions. If a rival is set on disproving your theories, you better be sure that there's no confirmation bias in your data and that your findings are reproducible.

But, like everything, competition also has a dark side. In science that can mean rushed decisions, a loss of focus, an emphasis on secrecy, and even bad behavior. Some scientists become so disenchanted with the process that they leave the field altogether. In a 2012 editorial in *Scientific American*, F. C. Fang and Arturo Casadevall wrote:

> As competition over reduced funding has increased markedly, these dis-advantages of the priority rule [the first who reports a finding gets the credit] may have begun to outweigh its benefits. Success rates for scientists applying for National Institutes of Health funding have recently reached an all-time

low. As a result, we have seen a steep rise in unhealthy competition among scientists, accompanied by a dramatic proliferation in the number of scientific publications retracted because of fraud or error. Recent scandals in science are reminiscent of the doping problems in sports, in which disproportionately rich rewards going to winners has fostered cheating.[19]

Looking back on my own experiences during *Deepwater Horizon*, it's not hard to see all of the mistakes I made. Some of these were the result of being in this competitive mindset. Other mistakes were the result of a lack of coordination among independent researchers, which was brought on by the chaos and collapsed time frame that accompanies a crisis event. In some ways, we were each trying to find Atlantis instead of contributing to solving the larger problems.

If I could have a redo, I never would have studied the plumes to begin with. I would have studied the surface oil instead. At the time, we didn't look at the surface oil because it wasn't sexy enough. We didn't think it was breaking any new ground—no one was going to get their name out there studying the composition of the slicks.

But in the end, the weathering and evaporation of all that surface oil *was* important. For me, it was more important than the plumes. It was a crucial component in the final calculation used to determine how much oil was left floating in the Gulf after the well was sealed. And the knowledge gained with regard to weathering and evaporation of surface oil is simply more transferable to other oil spill events. But as we'll explore in the next chapter, it took time to get there—Collin Ward's definitive research on photochemistry wasn't completed until eight years later. This, of course, is perfectly illustrative of how science works. No one answers questions definitively in a one-thousand-word news article or in a two-minute story on the radio. Science is all about the process. And in the heat of the moment, it can be truly difficult to distinguish what the best way to proceed really is.

While there was certainly competition during *Deepwater Horizon*—for funding, for research vessels, for prestige and recognition—at the same time, we have to acknowledge that scientists also rose to the challenge and proved our worth. We were nimble and ready to adapt to changing circumstances. We had already developed amazing, one-of-a-kind technologies, and these instruments helped us do things like take a sample of hot oil and gas from a deep-sea wellhead, where the pressure was a crushing 2,200 pounds per square inch, and bring it up to the surface without the sampling device exploding along the way. We were able to identify and map a twenty-two-mile-long plume of subsurface hydrocarbons in just three days, pinpointing the composition, direction of flow, and speed at which it was traveling.

The fact of the matter is that the deliverables from those first few months are awesome. The papers that were published in the years to come set the standard for oil spill research. With no advance warning and no planning, independent scientists were willing to drop everything in their lives and make use of their preexisting knowledge to do good science on the fly.

We made our share of mistakes, and many people were left feeling frustrated both during and after the spill. But at the same time, our value was recognized. The sample that we took from the well played an integral part in the US government's calculation that determined how many billions of dollars BP had to pay in fines and cleanup costs.

And yet, despite our contributions, somehow the worst of the clash between independent scientists and government responders was yet to come.

Notes

1 NIUST was a cooperative effort between the University of Mississippi (Oxford) and the University of Southern Mississippi (Hattiesburg). It was established by NOAA.

2 M. Schrope, "Oil Cruise Finds Deep-Sea Plume," *Nature* 465 (2010): 274, www.nature.com/articles/465274a.

3 One person who did expect this outcome was Scott Socolofsky, now at Texas A&M, whose graduate research at MIT (advised by Eric Adams) focused on the formation of subsurface plumes.

4 Justin Gillis, "Giant Plumes of Oil Are Forming Under the Gulf," *New York Times*, May 15, 2010, www.nytimes.com/2010/05/16/us/16oil.html.

5 Personal correspondence.

6 Once it became clear that the plumes were there to stay, the gag order was modified: don't tell the press we told you not to talk to them. See M. Schrope, "Deepwater Horizon: A Scientist at the Centre of the Spill," *Nature* 466 (2010): 680–84, https://doi.org/10.1038/466680a.

7 Schrope, "Oil Cruise Finds Deep-Sea Plume," 274.

8 "Scientists Find Huge Oil Plumes in Gulf," *NPR*, May 16, 2010, www.npr.org/templates/story/story.php?storyId=126870185. Emphasis mine.

9 "Huge Underwater Oil Plumes Found in Gulf of Mexico," *Associated Press*, May 16, 2010.

10 E. Pilkington, "Submerged Oil Plumes Suggest Gulf Spill Is Worse Than BP Claims," *Guardian*, May 16, 2010, www.theguardian.com/environment/2010/may/16/gulf-oil-spill-bp.

11 "Scientists Find Huge Oil Plumes in Gulf," *NPR*, May 16, 2010. Emphasis mine.

12 Safety was also a concern. We had to apply for permission to use the sampling device and carry out additional testing to ensure that it would not explode, as its purpose was to collect highly pressurized gas.

13 Eight hundred thousand barrels of oil were collected in this way. The collection vessels simply burned off the gas that came with it. See M. K. McNutt et al., "Review of flow rate estimates of the Deepwater Horizon oil spill," *PNAS* 109, no. 50 (December 20, 2011): https://doi.org/10.1073/pnas.1112139108.

14 S. Sylva, interview, "Science in a Time of Crisis: Sampling the Source," WHOI, 2011, www.whoi.edu/deepwaterhorizon/chapter3.html.

15 For the full account, see L. Lippsett, "Four Men. Twelve Hours. One Crucial Sample," *Oceanus Magazine*, July 21, 2011, www.whoi.edu/oilinocean/page.do?pid=53439&tid=282&cid=106909.

16 Ben Van Mooy, a molecular biogeochemist from WHOI, also analyzed the oxygen levels throughout the plume and determined that there was only a small dip when compared with the levels initially reported (which were measured with different

equipment). John Kessler followed up with a more comprehensive study over the course of the summer and determined that while oxygen levels did eventually drop by as much as 30 to 40%, there was thankfully not enough oxygen loss to result in any dead zones. See: M. Du and J. D. Kessler, "Assessment of the Spatial and Temporal Variability of Bulk Hydrocarbon Respiration Following the Deepwater Horizon Oil Spill," *Environmental Science Technology* 46, no. 19 (2012): https://doi.org/10.1021/es301363k.

17 Personal communication.
18 Personal communication.
19 Fang and Casadevall, "Winner Takes All," *Scientific American* 307, no. 2 (August 2012): 13, https://doi.org/10.1038/scientificamerican0812-13.

6   Piece by Piece

By the end of middle school, many children in the United States will have learned the basic principles of plate tectonics in science class. This unifying theory provides such a clear explanation for so many geological phenomena, from earthquakes to volcanic island chains to the formation of mountain ranges, ridges, and trenches, that it's hard to believe that it was not accepted into mainstream science until 1965.

As a matter of fact, none other than Albert Einstein wrote a foreword praising an alternative geological theory put forth in *The Earth's Shifting Crust* in 1958. Despite the name, this particular book, written by the American geologist Charles Hapgood, did *not* support the theory of continental drift.[1] Even though a German meteorologist had first proposed the idea in 1912, plate tectonics failed to gain acceptance for another fifty years—a lack of supporting evidence being one of the main challenges. Critics often take examples such as this one to argue that scientists are always changing their mind. If someone as smart as Einstein was skeptical of plate tectonics, then how can a regular person ever trust anything a scientist says? It could all change by next Tuesday!

But for anyone with a more realistic understanding of the scientific process, it's clear that this example is a perfect illustration of why science does work. Every scientist knows that science is iterative. Completing a scientific inquiry into a topic is like working on a jigsaw puzzle. Each new discovery is a single piece. You're not always going to put that piece in the right place within the puzzle. Maybe you'll have it rotated 180 degrees in the wrong direction. But eventually, as more and more evidence is gathered, and more and more pieces are filled in, you come to understand where and how your own particular piece fits together with everything else.

Sometimes, the picture that emerges is so radical—like the idea that the Earth is not the center of the universe, or that the Earth's surface consists of giant rocky plates that are moving independently across a hotter, more malleable layer—that it changes the accepted world view to such an extent that there is a paradigm shift. Society, and scientists, need time to process and accept such paradigm shifts. In some cases, we need lots of time. But as we all know, that's the beauty of science: it is not a collection of immutable facts written in stone. It is a self-correcting, never-ending search for the truth. If

DOI: 10.4324/9781003341871-9

there's one thing we can count on, it's not that scientists are constantly changing their minds. It's that our successive investigations into the same topic will gradually reveal a clearer picture of what we know—and what we don't. While this may sound frustrating to some, it is the very nature of this quest that makes science so exciting.

The Oil Budget Report

On July 15, 2010, BP announced they had successfully capped the leaking well. While the well wasn't declared officially sealed until September 19, the leak, for all intents and purposes, had finally been stopped, eighty-seven days after the explosion.

Even though the response team's primary goal had been attained, there was plenty of cleanup work to do. The oil that had already spilled was still floating in slicks across the Gulf, washing up on shorelines in the coastal states, and drifting in massive subsurface plumes. A thorough assessment of the environmental damage also needed to be completed.

As part of the latter task, roughly two weeks after the well was capped, on August 2, the National Incident Command released a mass balance report to answer the following questions: How much oil had spilled? And what happened to it? Although this brief three-page report, accompanied by an easy-to-read pie chart, was specifically designed to reduce anxiety regarding the spill's environmental impact, it immediately had the opposite effect: it became a lightning rod.

The basic assessment was the following. Over the course of eighty-seven days, approximately 4.9 million barrels of oil (210 million gallons) had spilled into the Gulf. Of this,

> it is estimated that burning, skimming and direct recovery from the well-head removed one quarter (25%) of the oil released from the wellhead. One quarter (25%) of the total oil naturally evaporated or dissolved, and just less than one quarter (24%) was dispersed (either naturally or as a result of operations) as microscopic droplets into Gulf waters. The residual amount—just over one quarter (26%)—is either on or just below the surface as light sheen and weathered tar balls, has washed ashore or been collected from the shore, or is buried in sand and sediments.[2]

If the government hadn't suffered from such a credibility problem throughout the *Deepwater Horizon* crisis, this might have been the end of the story. But instead, the reassuring tone resulted in an already skeptical public raising their eyebrows another inch higher. Independent scientists, meanwhile, parsed the language of the report.

What many people found particularly troubling was that the NIC report *appeared* to suggest that only 25% of the spilled oil remained in the Gulf ecosystem. In fact, the report doesn't state this at all. But the ambiguity—

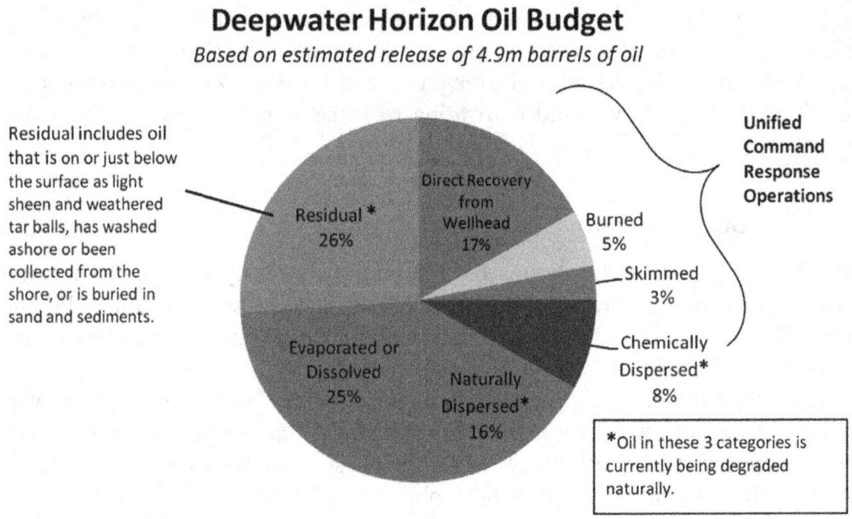
Deepwater Horizon Oil Budget
Based on estimated release of 4.9m barrels of oil
Residual includes oil that is on or just below the surface as light sheen and weathered tar balls, has washed ashore or been collected from the shore, or is buried in sand and sediments.
Residual * 26%
Direct Recovery from Wellhead 17%
Burned 5%
Skimmed 3%
Chemically Dispersed* 8%
Naturally Dispersed* 16%
Evaporated or Dissolved 25%
Unified Command Response Operations
*Oil in these 3 categories is currently being degraded naturally.

Figure 6.1 The National Oceanic and Atmospheric Administration's estimate of what happened to the oil. (J. Lubchenco et al., NOAA, 2010.)

intentional or not—was such that everyone who read the report, from politicians to journalists to members of the public, seemed to think that this was the main takeaway.

Reading closely, it is clear that the oil budget is stating that somewhere between 49 and 74% of the spill remained, depending on how you interpret the word *dissolved.* On page 2, the definition of "dispersed oil" is unequivocally stated:

> For the purpose of this analysis, 'dispersed oil' is defined as droplets that are less than 100 microns—about the diameter of a human hair. Oil droplets that are this small are neutrally buoyant and thus *remain in the water column* where they then begin to biodegrade.[3]

Dissolved oil was less explicitly defined: "Dissolution is different from dispersion. Dissolution is the process by which individual hydrocarbon molecules from the oil separate and dissolve into the water just as sugar can be dissolved in water." But as anyone who has ever added a spoonful of sugar to a cup of coffee knows, the sugar doesn't go away just because you can't see it. It's still there.

Unfortunately, most readers didn't seem to get past page one of the report: the framing of the words "24% was dispersed" and "25% of the oil … dissolved" was widely understood to mean that this oil was now gone, not floating in a diluted form in the water column. The majority of news outlets apparently failed to read the definitions of the terms on page 2 and completely misinterpreted the findings. Most suggested the NIC was trying to whitewash the outcome.

And from there, it all went downhill.

Backlash

Approximately two weeks after the NIC's oil budget was released, a team of independent scientists from the University of Georgia published their own report. The point of this report was not to dispute the NIC's assessment. It was to better explain the assessment to politicians, journalists, the public, and non-governmental organizations (NGOs).

In their own words:

> The findings of the report are being widely reported in the news media as suggesting that 75% of the oil is "gone" and only 25% remains. However, many independent scientists are interpreting the findings differently, with some suggesting that less than 10% is "gone" and up to 90% remains a threat to the ecosystem. ... Oil that the NIC report categorizes as Evaporated or Dissolved, Naturally Dispersed and Chemically Dispersed has been widely interpreted by the media to mean "gone" and no longer a threat to the ecosystem. However, this group [the Georgia Sea Grant] believes that most of the dissolved and dispersed forms of oil are still present and not necessarily harmless. *In order to better illustrate to the media, the public, community leaders and political decisionmakers* the current status of oil in the ecosystem, this group focused exclusively on oil that actually entered Gulf of Mexico waters, omitting from its consideration oil that was directly captured from the wellhead.[4]

Scientists recognized that there was massive confusion and tried to step in, in order to add clarity. The report goes on to estimate that in a best-case scenario, 10% of the spill was removed through human intervention (burning and skimming), 12% had evaporated, and 8% had degraded (eaten by microbes). So that left at least 70 to 79% that was still in the Gulf in one form or another.

If you remove the 17% of directly recovered oil from the NIC's equation, as the University of Georgia did, you'll see that both reports wind up with roughly similar figures.[5] Are they the same? Not exactly, because the University of Georgia factored in a more precise number for the amount they believed had evaporated (7 to 12%), rather than lumping evaporation and dissolution together.

For me, this was a nonstory. We question and challenge each other all the time. That's how the process works. While the exact amount of residual oil was not the same in the two reports, it wasn't hugely different. The main difference was in the packaging.

Still, the fact that academics were stepping in to clarify the facts shouldn't have been a problem. But suddenly the big story had changed: now it wasn't that it *looked like* the government was trying to pull a fast one. Now, independent scientists had confirmed the government *was* pulling a fast one. Ledes such as the following were commonplace: "A group of scientists say that most of that BP oil the government claimed was gone from the Gulf of Mexico is actually still there."[6]

But who was pulling a fast one here? Did the media's reading comprehension skills suddenly plummet to middle school levels? Or in the rush to get a juicy, conflict-laden story out first, did news agencies simply overlook the finer details of what the independent report actually said?

Assigning blame to any one group in this situation is risky business. We all know what happens when someone suggests that the media is in the business of disseminating fake news. Or that the government is in the business of deceiving its citizens. But as the *Deepwater* example shows, declining confidence in public institutions, which reached dangerous new levels during the COVID-19 pandemic, had already been a significant trend in the United States for quite some time. Previous natural disasters such as Hurricane Katrina, where the public perceived the government response as a failure, only compounded the sentiment.

In all likelihood, no one was actively trying to deceive anyone. Each stakeholder group was simply trying to achieve their goals. Responders and industry wanted to calm an anxious public—and declare success. The media wanted to present what it thought were different points of view—and attract an audience. Environmental activists wanted to be sure that the full extent of the damage was not being overlooked—and enact policy change. Politicians wanted to obtain results for angry and skeptical constituents—and get reelected. And scientists wanted to collect more data to help complete the puzzle—and add rigor and continue to expand science. Everyone had competing goals, as well as competing time frames. Mistakes were certainly made. But we should have been able to overcome them.

Box 6.1 Dude, You're Speaking Romulan!

As the environmental chemist on the *Endeavor*'s plume-finding research team, I often wondered aloud how we could exploit the aqueous solubilities of the petroleum hydrocarbons—benzene, toluene, ethylbenzene, and total xylenes—to understand plume formation. I suspected the key to knowledge lay in the plume's chemical properties.

"Dude, you're speaking Romulan!" one of my colleagues finally blurted out. The engineers in the group gave me a look, and steered the conversation to the relative merits of different types of statistical processing of data collected in and around the plume. I don't know about statistical processing, so I hit back: "Dude, *you* are speaking Romulan."

As *Star Trek* fans know, Romulans are a race often at odds with the Federation. Romulans speak in three dialects and write with square or rectangular letters. Telling your colleague that they are speaking Romulan is a friendly way of saying, "I don't understand you." Either you are using jargon, speaking too fast, using acronyms, or jumping over the natural progression of an argument or idea.

What is surprising is that we have these communication breakdowns despite my colleagues also being my friends. We work at the same institution. I have been to sea with them. I know their dogs, eat dinner at their homes, and jointly lament the standing of the Red Sox. Even though we know each other well, our differing scientific specializations can cause us to speak different languages. For us, our small group was willing to recognize these differences and set the ground rules for using the "Romulan phrase."

Consider how different scientists might discuss the usage and effects of dispersants on oil slicks. A physical chemist will discuss the forces that create small droplets of oil out of large slicks; an oil spill scientist will argue the trade-offs of using dispersants to protect coastlines and wildlife versus increasing the oil content below the ocean surface; a microbiologist will explain how small droplets of oil created by dispersants are more available to microbes to degrade; and a toxicologist will explain how dispersed oil increases toxicity to aquatic life. What would happen if a layperson, member of the media, or policymaker received just one of these responses, or two?

Since the early 2000s, there has been a major push to train scientists to communicate with the public, the media, and policymakers. Programs around the country and at many universities have workshops or courses for scientists and students. I coteach a graduate course with my close friend, the journalist Lonny Lippsett, entitled "How Not to Write for Peer-reviewed Journals: Talking to Everyone Else," and often beat the drum for improving science communication.[7] One year, at our first class, we asked students to briefly describe their research. I can't say I fully understood what everyone said.

Despite the stereotypes on TV and in movies, scientists have a complex and competitive culture. We aspire to be a "peer," to learn the language, and to use that language to communicate and defend our science among learned peers. We tend to forget—or ignore—what it's like not to understand. We go on autopilot.

So before teaching scientists how to speak to nonscientists, perhaps scientists should first learn how to speak to other scientists. Donald Kennedy, former editor of *Science* and president emeritus of Stanford University, mentioned this very problem in an editorial in that magazine in 2007. He wrote, "It's clear that accessibility is a problem, because we're all laypeople these days: Each specialty has focused in to a point at which even the occupants of neighboring [scientific] fields have trouble understanding each other's papers."

Almost every pressing scientific and environmental problem demands the attention of scientists from diverse disciplines as well as the expertise of economists, planners, and sociologists. With a little effort and less ego, we need to aim for a lingua franca that can be understood by a politician, a shrimp farmer, a toxicologist, a lawyer, an accountant, and a Romulan, too.[8]

In the midst of all the backlash over the University of Georgia report, our research on subsurface plumes was published in *Science*. As the first peer-reviewed study on *Deepwater Horizon*, this was a huge, validating moment for all of us. On August 19, together with Woods Hole Oceanographic Institution (WHOI) vice president Rob Munier, lead author Rich Camilli, and the other coauthors, we held a press conference where we announced our methodology and findings. We emphasized that there was still a lot to learn about the plumes, including their toxicity and the rate at which they would biodegrade.

The speed at which we had obtained funding, organized and conducted the research, and had our results go through peer review was so fast that it felt like a victory not only for us, but for science in general. This was some of the most cutting-edge work that had been conducted during an environmental crisis, and we thought we were adding more clarity to a situation that was still misunderstood. We essentially said, "Yes, the dispersed oil is still in the water column. But it's not a river of pure oil as many people are imagining. It only consists of certain compounds. And it's not causing these giant dead zones that everyone was afraid of."

Interestingly, this moment wound up sticking with me for another reason. It was something that Thad Allen, the Incident Commander, said to me in the weeks following our press conference: "I don't understand why everyone is so excited. The well was sealed last month. You guys are too late."

Thad and I have a good relationship. He wasn't saying this to spite me, or because he was feeling the heat of the oil budget criticism and was frustrated with scientists in general. He was just being honest. It is difficult for many nonscientists to appreciate just how big a deal having a paper published in *Science* is—in addition to it being the first paper on the subject. It is also hard for nonscientists to appreciate that there is long-term value in what we did. It was this moment, along with several others in 2010, when I fully realized that most people simply do not understand our culture and what our metrics of success are. It turns out that part of effective science communication is not only delivering your message, but also explaining the scientific process and how we define success to other stakeholders.

In Thad's mind, he was fighting a fire, and now that fire was out. And here we were, writing a paper about how hot the fire had been two months ago. What was the point of holding a press conference?

Us versus Them

Sure enough, the intended message of our press conference did not go through. Much to our horror, our work on subsurface plumes got swept up in the ongoing "independent scientists versus the government" narrative. Repeated news stories in the days that followed placed us squarely in the anti-government camp. Quotes like the following were all too common:

> Earlier this month, top federal officials declared the oil in the spill was mostly "gone," and it is gone in the sense you can't see it. But the chemical ingredients of the oil persist more than a half-mile beneath the surface, researchers found.[9]

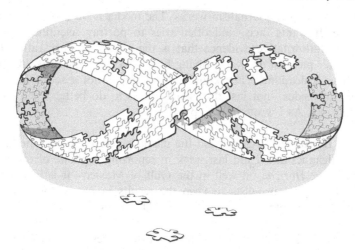

Science is an infinite puzzle.

Figure 6.2 One major disconnect between scientists and nonscientists is the realization and acceptance that science is rarely resolved. It is self-correcting, incremental, and ever-expanding in a nonlinear fashion.

> Earlier this month, a report from the National Oceanic and Atmospheric Association said that almost 75 percent of oil from the well had been captured, burned, naturally biodegraded or dispersed in the water, and that the dispersed oil was biodegrading quickly. But this week, two reports from outside scientists have taken issue with the government's conclusions.[10]

Such stories incensed both the responders and the WHOI team. Occasionally, another scientist would add more fuel to the fire with an unfounded speculation, such as: "We absolutely should be concerned that this material is drifting around for who knows how long. They say months in the [research] paper, but more likely we'll be able to track this stuff for years."[11]

In the heat of the moment, I wrote an op-ed in an attempt to correct the narrative. Titled "How Reporters Mangle Science on Gulf Oil," the piece went live on *CNN* on August 25. Looking back, I know that everyone was at fault for the way things transpired: the government, journalists, industry, NGOs, and scientists. But more than anything, there was a complete disconnect with regard to our timescales. The media's mission is to provide immediate, definitive information about unfolding events to an anxious public. But even as I write this eleven years later, the numbers in the oil budget are *still* being refined. Imagine telling a reporter: "Get back to us in a decade or two. By then we should have a better—but not definitive—idea of what happened to all that spilled oil."

That's not the way journalism works. The media is on a twenty-four-hour news cycle. It wants facts, and often tries to portray scientific findings as conclusive, authoritative evidence that is the end of a story rather than a much smaller piece of an ever-evolving bigger picture. Scientists, in contrast, do not have a deadline. They know that science is not a set of facts, but a method. Both sides—not just the media—need to do better when trying to bridge this gap. As I wrote in my op-ed:

When researchers present what the media perceive as "big" findings—as my colleagues and I did last week in reporting a plume of oil from the *Deepwater Horizon* oil well in the Gulf of Mexico—it is incumbent on scientists and journalists to keep the results in perspective and refrain from veering into misleading waters. Unfortunately, in this case, both parties failed.

Reporters and editors, in their quest for the biggest story possible, injected their reports with implications unintended by scientists. For their part, scientists from various corners of government and academia—including our group at the Woods Hole Oceanographic Institution (WHOI)—let it happen. In some cases, they may have even encouraged it. ...

I must have spoken with at least 25 journalists last week, and despite my every effort to explain our findings, the media were more interested in using the new information to portray a duel between competing scientists. The story turned into an us-versus-them scenario in which some scientists are right and others are wrong. Seeking to elucidate, I felt caught in a crossfire. ...

Our research confirmed the existence of a subsurface oil plume in June that did not come from a natural sea floor oil seep and that was not substantially degraded by deep-sea microbes. The research added new information to an unfolding investigation, but the media seemed more interested in whether our work decided whether NOAA or the Georgia group was right.

Even though my colleagues and I repeatedly avoided contrasting our results with previous NOAA estimates that were interpreted to mean some 75 percent of the spilled oil was already gone from the Gulf, much of last week's coverage of our work made that a prominent part of the story.

For example, the *Washington Post* reported, "Academic scientists are challenging the Obama administration's assertion that most of BP's oil in the Gulf of Mexico is either gone or rapidly disappearing—with one group Thursday announcing the discovery of a 22-mile 'plume' of oil that shows little sign of vanishing."

In doing so, it cast our results as evidence of sorts that the NOAA estimates were wrong, and at the same time had the effect of giving the Georgia work our imprimatur. Neither of these conclusions was ever meant to be drawn from our research on the oil plume. This reasoning implicit in the media coverage was not only premature, but it might turn out to be wrong.

Science does not work that way. It is incremental. It is not a house of cards where one dissenting view leads to a complete collapse. Rather, science is more like a jigsaw puzzle. Each piece is added. Occasionally a wrong piece may be placed, but eventually science will correct it.

Both the corrections and the completion of any scientific puzzle take time.

Scientific peers regulate the process of presenting hypotheses, acquiring data, and assessing them. In this process, questions are asked, gaps get filled, inconsistencies are hammered out, discoveries are made, problems get solved, and knowledge is obtained. Science's regulatory systems have a very solid record of accomplishment.

Unfortunately, the process takes months or years, but in this case, it has been compressed into days with dueling reports and news conferences on the fantails of boats. News organizations haven't the luxury of time to distill scientific findings and put them into context, which increases the risk of oversimplifying scientific findings.

Some of these problems are scientists' fault. In our world in the peer-review process, we liberally, passionately, sometimes harshly interrogate each other, where we argue over details and interpretations of research results to ensure that they are bulletproof. Out of the academic world, reporters can magnify negative comments by scientists about research results.

As the number of science journalists gets smaller, this problem will grow. One solution is for scientists to gain skills needed to bridge the communication gaps between the academic world and the lay public, media and policymakers.

In addition, scientists need to learn how to say "no" to reporters. For many of us, we desperately want to please a reporter, who for the first time cares about what you do. And scientists, including me, have egos, so we want our thoughts and work recognized.

But scientists have a better chance of getting the story straight if they listen carefully to the questions asked by reporters and understand the reporters' goals.

In 1994, 11 scientists published a study, "The fate of the oil spilled from the *Exxon Valdez*: The mass balance is the most complete and accurate of any major oil spill." Of these 11 authors, six were NOAA scientists, one was from academia and four from four different consulting firms. The *Exxon Valdez* oil spill happened in 1989.

Science takes time. If it took five years to "balance the books" on how much oil was spilled and where it went for the *Exxon Valdez* spill, how are we getting estimates of the *Deepwater Horizon* spill only weeks afterward? It's not trivial to decipher something as vast, fluid, complex, and inaccessible as the ocean.

So given that it is so early in this investigation of the *Deepwater Horizon* oil spill, I would consider both the NOAA and Georgia studies as first passes. Neither is absolutely right or wrong. They are certainly not the definitive findings, but should be thought of as a foundation from which to work, road maps to use in assigning future research assets in examining the transport and fate of oil in the Gulf of Mexico.

Those road maps will be refined into robust values as more information becomes available. Eventually, teams of scientists will be able to "balance the books" for the *Deepwater Horizon* spill, too.

Over the next few months, many scientific studies on the spill will be published and reported on. Journalism, the first draft of history, is incremental, too. Consider each scientific report like a chapter in an epic novel, and not necessarily in order. Let the dust settle and read the book in a few years.[12]

So now, over a decade down the road, what *do* we know about the oil budget? As I mentioned in the previous chapter, one of the biggest outcomes from all of the *Deepwater Horizon* research was the discovery that photochemistry, or the impact of sunlight, plays an important role in degrading oil slicks—and the most significant degradation happens within the first week. This was new information to all of us, and we didn't develop this hypothesis until 2012. And it wasn't until 2018 that Collin Ward and I, along with eight other colleagues, published a peer-reviewed paper on the findings.[13] Although it took a long time to confirm, this was a big puzzle piece, because we now know that the earliest estimates on evaporation were way off.

In the meantime, a lot of other models on the oil budget have been published, but each one is different because it draws on a different study, which uses different terminology and different metrics from the others. Up until now, all of these different results haven't funneled in a comprehensive way. The spill was so big, and the ocean is so complex, that trying to determine the mass balance is a bit like trying to figure out what the daily weather was like for every day in the spring and summer of 2010, across the entirety of the United States, but only using data from ten disparate weather stations.

Because of this, the results are never going to be comprehensive. But they are certainly going to be more accurate, and more multifaceted, than the first two estimates the NIC and the University of Georgia released. Ultimately, science did, and is still doing, its job.

But as far as the affected public was concerned, we might have been too late.

Notes

1 Hapgood's theory, known as crust displacement, was that the large mass of ice during the last ice age caused the earth's crust to "slip" as a unified whole over the mantle, thus explaining how parts of Antarctica remained ice free until recently. Einstein wrote, "I frequently receive communications from people who wish to consult me concerning their unpublished ideas. It goes without saying that these ideas are very seldom possessed of scientific validity. The very first communication, however, that I received from Mr. Hapgood electrified me. His idea is original, of

great simplicity, and—if it continues to prove itself—of great importance to everything that is related to the history of the earth's surface" (Einstein, May 18, 1954, Einstein Archives).

2 Lubchenco et al., "BP Deepwater Horizon Oil Budget: What Happened to the Oil?" NOAA (2010): https://repository.library.noaa.gov/view/noaa/19.

3 Lubchenco et al., "BP Deepwater Horizon Oil Budget." Emphasis mine.

4 C. Hopkinson et al., "Outcome/Guidance from Georgia Sea Grant Program: Current Status of BP Oil Spill," University of Georgia, August 17, 2010. http://oceanleadership.org/wp-content/uploads/2010/07/GeorgiaSeaGrant_OilSpillReport8-16.pdf. Emphasis mine.

5 Thus, instead of using a total of 4.9 million barrels of oil, as the NIC did, you use a total of 4.1 million barrels (17% = 0.8 million).

6 "Gulf Oil Spill: University Study Contradicts Government Estimates," Associated Press, August 17, 2010.

7 I owe a tremendous amount of gratitude to Lonny Lippsett whose inspiration and dedication to science communication left an indelible mark on me. I might have been listed as a "coteacher," but Lonny was the real instructor in the class.

8 This article was originally published in *Eos*: C. Reddy, "Dude, You're Speaking Romulan!," *Eos* 91, no. 42 (October 2010): 384.

9 "Major Study Proves Oil Plume That's Not Going Away," Associated Press, August 19, 2010. At no point does the NIC report use the word "gone."

10 "Gulf Oil Plume Map Adds to Debate Over Spill's Undersea Impact," *PBS*, August 19, 2010.

11 "Major Study," Associated Press, August 19, 2010.

12 This article was originally published on CNN: C. Reddy, "How Reporters Mangle Science on Gulf Oil," *CNN*, August 25, 2010, www.cnn.com/2010/OPINION/08/25/reddy.science.media/index.html. Permission to reprint courtesy CNN.

13 C. P. Ward et al., "Partial Photochemical Oxidation Was a Dominant Fate of Deepwater Horizon Surface Oil," *Environmental Science & Technology* 52, no. 4 (2018): 1797–805.

7 Countering Scientific Misinformation

In 2011, about one year after the *Deepwater Horizon* spill, I was out collecting samples on Elmer's Island, Louisiana, with a colleague. Anyone who has ever done fieldwork in a public area knows that it is pretty clear to others what you're up to. We might not have been dressed in white lab coats, and no one had stuck a KICK ME sign on my back, but the purple gloves and lab notebook certainly shouted "science world!" to the general public. And that made us an attraction of sorts, because a lot of people still wanted to talk about the spill—especially with a so-called expert.

On this particular day, a woman came up to us and started to chat. We were busy collecting little oil-and-sand patties—about the size of a quarter—that had formed after the oil had washed up on the beach. These patties persisted for such a long time that we were flying down to the Gulf on a regular basis between 2011 and 2015 in order to learn more about how the oil was breaking down. Because we were so focused on our task, we didn't completely engage with the woman at first.

But, pretty quickly, it became clear that she was extremely upset. At one point, she stuck out her forearm and pointed to a tiny red spot and said, "Look at this! I never had this spot before *Deepwater Horizon*. And now I do, and it keeps getting bigger!" Now, I was not going to dispute what she was saying. But at the same time, I am not a medical doctor, and I had no idea what she was showing me.

She, on the other hand, was pretty sure what had caused the spot: dispersants. The alleged toxicity of dispersants—which responders applied to break up the oil both on the ocean surface and at the wellhead—was one of the big media stories early on, and it stuck with the affected public. In the years that followed, whenever we would drive west along Interstate 10 from Pensacola, Florida, stopping at our different sample sites along the way, we would pass hundreds of billboards. And they all advertised the same thing: class-action lawsuits. For some people, these law firms were reinforcing the idea that the dispersants had made locals sick; for others, the billboards may have actually been putting the idea in their heads. Either way, they weren't helping anyone.

Dispersants are a known quantity in the oil-spill response toolbox. While they are chemicals, they are relatively less toxic than the oil itself. And in

DOI: 10.4324/9781003341871-10

Deepwater Horizon, the ratio of dispersants to petroleum was 1 to 20. So when you're talking about pollutants and toxicity, it's clear which one was going to have a bigger effect.

After we heard the woman say, "This spot is from the dispersants," we both thought, "If this woman wasn't a responder, then she wouldn't have been near any dispersants in the first place. We should tell her that she doesn't need to worry about that particular issue." So my colleague, who is a very nice guy, and not arrogant or condescending in the least, said, "You know, dispersants are not that big a deal for human exposure when compared to the oil itself."[1]

Even though we thought we were trying to soothe her anxiety, this statement so enraged the woman that she actually chased after us right there on the beach. Afterward, we laughed about it—"Holy shit, I can't believe we almost got beat up while collecting samples!"—but we weren't laughing at her. It was the shock of what had happened, and just how badly we had misjudged the situation.

Looking back, I now understand that this woman was not interested in what was right and what was wrong. She just wanted someone to listen to her. Someone to say, "Hey, I'm sorry you had to live through that. I'm sorry that you don't feel well. We care about you." But in telling her that the dispersants weren't a problem, we were reinforcing the establishment's message, which she had already decided was a lie. And that meant we were saying we didn't care, even though we intended to communicate the opposite.

After this, my default reaction whenever anyone approached me in public was to simply say, "Hey, I'm only a chemist. Science takes a long time. This isn't like the television show *CSI*, where everything gets solved in fifty-nine minutes. Right now, I'm just collecting samples in order to analyze them. Once I have some results, I'd be happy to share them with you." If it seemed like they were open to discussion—and many people were—we would chat, and I would give them my email address. But if not, I didn't force the issue. I would simply listen to what they wanted to tell me and validate their feelings and move on.

What is the best approach when confronted with misinformation? There is no one-style-fits-all, because it often depends on the context. What is certain is that some misinformation—like the idea that dispersants or vaccinations do more harm than good, or that global warming is a conspiracy—does cause real harm to society. We know, of course, that science isn't perfect. During COVID-19, some people have had terrible reactions to the vaccines.

And yet, it is so easy to become frustrated when people choose to deny clear evidence that, for example, vaccines provide a net benefit to society. It is in our own interest to counter such scientific misinformation. But while it's tempting to believe this is a problem of scientific literacy or a lack of access to the evidence, trying to educate someone else—or, as they might see it, force your own worldview on them—will rarely work. They've already made up their mind. The only thing you're likely to accomplish is that you'll make an enemy.

Box 7.1 Did the Dispersants Work?

There are several parallels between the public's skepticism of dispersants during *Deepwater Horizon* and vaccine hesitancy during COVID-19. But there is also one important difference: during *Deepwater Horizon*, there was no well-designed comprehensive field study that monitored its efficacy. Today, when someone refutes the effectiveness of the COVID-19 vaccines, people can go look at the data for themselves and say, "Oh, it really did have a beneficial impact."

But during *Deepwater Horizon*, that didn't happen. In the years that followed the spill, I gave over seventy public talks. And after each lecture, the first question from the audience was inevitably: did the dispersants work? It wasn't just members of the affected public. My colleagues, too, were curious.

And because we didn't have any robust field data, I couldn't give a definite answer. So my response was always: at the very beginning of the spill, when everything was going sideways, when there were sixty thousand barrels of oil coming out of the sea floor per day, and the air quality along the coast was plummeting because of all the oil compounds floating around, and the responders were worried about all these other variables from wildlife to fisheries, someone said, "Hey, if we add dispersants directly to the wellhead, we think the oil won't clump up as much, and we'll also improve the air quality. It's better to have some of the most volatile oil compounds, like benzene, diluted underwater rather than on the surface."

The important thing was that the people who made that decision were the right people to do so. Their conclusions might have been wrong—for the record, I don't think they were—but they were the ones who had the most training, who had done the most research, and who had twenty-five years' experience of working with dispersants and oil spills. This wasn't some half-baked solution that the government came up with at the last minute. These were the leaders in the field.

Unfortunately, where they did make a mistake was in not collecting data. If they had developed a robust model and air-quality monitoring network at the very beginning, they could have said, "We're going to look closely at how things change once the dispersants are added. We're going to study this outcome in such depth that it's going to be a textbook case as to whether underwater dispersants work or not. We're going to see exactly how much the air quality improves."

But because everything was moving so fast, that's not what happened. What's more, the response team was unprepared to deal with such intense public skepticism.[2] So when people did push back on the dispersants, no one was able to refute the claims with authority, because there just wasn't enough real-time data.

Cognitive Bias

In order to better understand how we can effectively counter the spread of scientific misinformation, let's start at the beginning. How does such misinformation spread? In today's society, and particularly during the COVID-19 pandemic, social media bore the brunt of the blame for the dramatic rise of misinformation. But it's important to remember that the origin of such viewpoints is not on platforms like Twitter and Facebook. Such ideas come from people talking among themselves. Social media is merely an amplifier that allows misinformation to spread faster, with a further reach, and more polarized outcomes.

The onset of a crisis event is often marked by high levels of uncertainty. The public is hungry for information that scientists are unable to provide. *How long will COVID last? How infectious is it? What is the fatality rate? Will it disappear in the summer? Are there any long-term effects? Is there a cure?* Search your memory, and you'll probably find that you yourself were wondering these same questions at the beginning of the pandemic.

As a novel virus, this was the first time the world was getting a glimpse of how humans responded to COVID-19. There was no way for any medical researcher to answer these questions with certainty in the pandemic's earliest days. But existing in a perpetual state of the unknown does not sit well with the human brain. When faced with a threat, we want immediate solutions. And if there is no reliable authority to provide the answers to our questions, we do the next-best thing: we come to our own conclusions, based on our own set of cognitive biases.

In a 2020 article in *Scientific American*, three such biases were identified as playing a critical role in the spread of COVID-19 misinformation: (1) an overreliance on information from people we trust, particularly when faced with information overload; (2) a tendency to focus on perceived threats and share information about them; (3) a tendency to look for patterns or ideas that match with our previous experiences or fit in with our knowledge base.[3]

The information vacuum at the beginning of a crisis is thus the perfect environment for the growth of conspiracy theories with ready-made answers, the targeting of the "other team" or "outsiders" as ready-made villains, and even well-intentioned but off-base speculations with no supporting evidence. In the long run, science will catch up, but, like the regeneration of a forest following a wildfire, it takes time for the slow-growing trees to shade out the faster-growing shrubs and grasses.

I would also argue that there is another variable that augments these cognitive biases. And it is this: certain populations are primed to view the authorities as inherently untrustworthy. In the case of *Deepwater Horizon*, the government had already lost most of its credibility because of the failed Hurricane Katrina response.

COVID-19 is a global pandemic, and therefore considerably more complex, but similar patterns have been observed. Historically, American medical

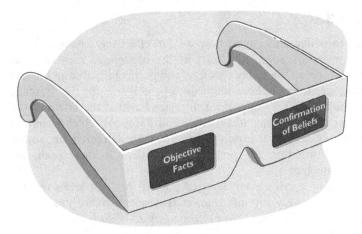

Figure 7.1 Overcoming cognitive bias is a major hurdle when communicating science. Being simply aware of the facts and beliefs is the first step to overcoming them.

institutions and researchers repeatedly used unethical means to exploit the bodies of African Americans without their consent.[4] This, combined with a lack of easy access to vaccination sites, accounted for the African American community having a larger percentage of unvaccinated individuals than other racial groups.[5]

The political divide, often viewed through the prism of rural versus urban America, is another example of how mistrust came into play during the pandemic. In October 2021, the COVID-19 death rate was three times higher in red counties than in blue counties, the largest observed gap since the introduction of the vaccines. As the *New York Times* wrote:

> This situation is a tragedy, in which irrational fears about vaccine side effects have overwhelmed rational fears about a deadly virus. It stems from disinformation—promoted by right-wing media, like Rupert Murdoch's Fox News, the Sinclair Broadcast Group and online sources—that preys on the distrust that results from stagnant living standards.[6]

When it comes down to it, all people, regardless of demographics, are going to be naturally skeptical of most proposed solutions during a crisis, because these solutions will never be perfect. There will always be a trade-off of some sort. Adding an unknown ingredient that is complex and perceived to be bad—like a chemical dispersant, or a vaccine that is known to have powerful side effects—is always going to be a tough sell. It may seem to the general public that scientists are taking a bad situation and making it worse.

But when you factor in a high level of mistrust of the authorities on top of that natural skepticism, what was already a tough sell may verge on the impossible.

Box 7.2 What the Experts Say: Masters of Messaging

When considering how to deal with misinformation, consider taking a page from the playbook of one of the experts: the nonprofit organization the Ad Council, founded in 1942. With an impressive track record of public service messages that range from increasing polio inoculations to wearing seat belts to AIDS education and prevention, Ad Council slogans have been a part of American life for eighty years.

Their latest campaign? Boosting vaccination rates during COVID-19. Drawing on their previous expertise, the nonprofit organization reminded everyone that "it's normal to have questions about the vaccines. Over and over again, [we have seen] that people do not want to be lectured or feel put on the spot—effective messaging must meet people where they are."[7]

This is a good reminder: while it's tempting for those of us in the sciences to rely solely on fact-based arguments to make a point, peer-reviewed studies have borne out a different truth. As the Ad Council stated, the way in which those facts are shared makes all the difference.

In a 2022 study, Nicholas Light and his colleagues noted the following: because "most extreme opponents of the scientific consensus tend to be those who are most overconfident in their knowledge, fact-based educational interventions are less likely to be effective."[8]

More successful approaches are thus participatory and grounded in "constant dialogue, exchange of knowledge and negotiation," and narrative.[9] Whatever your approach, it's important to recognize that a 100% success rate is unlikely. Professor Baruch Fischhoff put it best when he said: "The goal of science communication is not agreement, but fewer, better disagreements."[10]

The Integrated Fukushima Ocean Radionuclide Monitoring Project

In 2011, a 9.0-magnitude earthquake struck off the east coast of Japan, with a force so powerful that it shifted the earth's axis by roughly six and a half inches. The resulting tsunami that smashed into Honshu Island left over 18,500 people dead and flooded the Fukushima Dai-ichi nuclear power plant, knocking out the backup generators. With no power for cooling, the plant experienced a meltdown in three of its six reactors, which subsequently led to the discharge of a massive amount of radioactive water into the Pacific Ocean.

It was a mega disaster on many fronts: it was not one crisis event but three, and each one was inextricably tied to the other. On the scale of nuclear catastrophes alone, only Chernobyl was worse. And yet, because of the size of the ocean and the strength of the local currents, most of the radioactive particles were quickly swept out to sea and diluted in the seawater. In this sense, it seemed that Japan had dodged at least one bullet.

But just how diluted is safe enough? This has been an ongoing question in the decade since, because contaminated water continues to leak from the site even today. In 2014, it was announced that radioactive water from the meltdown was going to start washing up on North America's Pacific coast. Americans, understandably, began to worry: will it be safe to eat seafood? Will it be safe to go into the water?

One Canadian chemist and oceanographer, Jay Cullen of the University of Victoria, was so inundated with questions about the safety of the coastal waters in British Columbia that he decided to set up a radionuclide monitoring project, called the InFORM Network. The idea was to take the measured levels of radioactivity from multiple test sites and make them easily accessible to the public. And there wound up being good news to share: the peak levels of radioactivity measured in the water offshore were only one one-thousandth of the level allowed in Canada's drinking water.

Unfortunately, Jay discovered that sometimes people don't want to hear what science has to say. As he stated in a 2018 television interview,

> Radiation is a difficult thing to talk about because we can't see it, we can't taste it, and we can't smell it. The history of nuclear weapons and the fear that they impose upon people, and the fact that government and industry are all involved in nuclear technology—for a lot of people, that combination leads to distrust, and fear, and ultimately misunderstanding. ... One of our motivations [in setting up this website] was to counteract that misunderstanding. ... [Unfortunately,] simply by being a scientist and working at a university, that disqualifies me as a source of information for some people."[11]

But the real eye-opening moment for Jay wasn't that some people disputed the data. He knows that many people feel strongly about nuclear energy and expected that to be the case. What caught him off guard was that some people went after him by launching a hate campaign. He received death threats. His credentials as a scientist were questioned. He was called a shill for the nuclear industry.[12]

Jay is an outstanding individual. In the face of these attacks, he didn't hide below the parapet and wait for the controversy to go away. Instead, he continued to post updates of the latest measurements. Jay's attitude, and strategy, is one solution to countering misinformation. He believes that if you don't put the evidence out there, you are leaving behind a vacuum similar to the one that emerges at the beginning of a crisis. And without any evidence to the contrary, that vacuum will naturally fill up with falsehoods that lead to fear and degrading levels of mistrust in science.

Box 7.3 Logical Fallacies: The Appeal to Accomplishment

I follow a couple of doctors and medical researchers on TikTok. And in general, they are very well spoken, particularly when talking about COVID-19. However, there is one major mistake that they all seem to make. Whenever they are engaged in a heated debate with someone and aren't making headway, they always wind up rolling out their credentials. They usually say something like, "Okay, do you want to listen to this guy, who is not a doctor, or do you want to listen to me? I did a double major in chemistry and physics at Penn, went to med school at Harvard, and was a resident at Johns Hopkins."

Listing your credentials may be common practice in some situations, but trying to flex on people who are already skeptical of your message? It just doesn't work. All you do is wind up sounding like an elitist jerk. This type of unforced error, where you dismiss another person's point of view simply because they don't have the same accomplishments as you do, is known as the credentials fallacy. You aren't addressing what the other person has to say. Instead, you are telling them, "Your ideas have no merit because of who you are."

Every time I have heard someone list their degrees, it has always been when they are at their worst. You can see the frustration building on their face. Thus, it bears remembering, if you are debating someone, stick to the issue at hand. Some people are not going to listen to you no matter what. That's okay. If you sense that the conversation isn't going anywhere, simply repeat your main point and disengage. Don't take it personally and begin insulting the other person, because that's when you start to lose the people in the middle—the people who might actually be listening to what you have to say.

Social Media: Making a Positive Impact

Numerous studies have been done on social media, and by now, many of us know that the algorithms that power these platforms are able to leverage our cognitive biases in the worst possible way. The tendency to listen only to the opinions of like-minded people we trust? To share information about perceived threats? To look for patterns and ideas that reinforce our worldviews? To find an external villain to blame for life's assorted ills? Yep, that about sums it up.

Would vaccination hesitancy have become so widespread in the United States without social media? Perhaps not. But it is important to remember that it would have still existed. Understanding the subtle ways in which social media manipulates the human brain is important, but it doesn't mean you should shut it out of your life. After all, it's just a tool. And it's clearly a powerful one at that.

In the Introduction to this book, I gave an example of how one tweet I posted led to a positive outcome. In under 280 characters, I refuted the misconception that a broken gas pipeline in the Gulf of Mexico was another *Deepwater Horizon*. This single tweet caught the attention of a journalist at *Wired*, and she subsequently reached out to me and went on to write a more in-depth article on the event. In this way, I was able to communicate my message to a larger audience.

There are other productive ways for scientists to use social media. For instance, you can more effectively promote your research or publications to those who are in the same field. You can share what life in a lab or in the field is really like via photos and videos, helping to make science "cool" among the general public and possibly inspiring the next generation. Social media is a fantastic way to dispel mainstream stereotypes about scientists—that we're all super geeky, unathletic, and wear glasses—and be your genuine self at the same time.

It may also help you improve your fundamental storytelling skills: distilling your research or message into a format that a broader audience can grasp takes practice. But in some ways, a post on Instagram, circulated among your friends, may be less stressful than writing an op-ed in the local paper. If you bomb a story or a tweet on social media, you can always delete it and try again the next day. Eventually, through trial and error, you'll see what sort of content people respond to in a positive manner, and what sort of content is completely ignored.

Countering scientific falsehoods online may also seem like a good reason to get involved on social media. However, this is an area where it pays to be careful, because the nuances of human interaction are often lost in non-face-to-face encounters. At the beginning of this chapter, I told a story about a woman who was convinced that dispersants were the cause of her health problems. What would this interaction have looked like if it happened on Facebook or Twitter? Would I have recognized the depths of her despair and rage? Would I have realized that replying to her emotional distress with hard facts was not the right approach? Probably not.

In this particular case, even though we were face to face, we still blew it. But the intensity of this event left its mark, and I certainly learned a lot about what not to do in such a situation. When someone is upset, it is not the time to roll out your credentials or to explain why a person's beliefs are misguided. You may know that, scientifically speaking, they are misinformed. Over the course of collecting samples in the years after *Deepwater Horizon*, I had multiple people come up to me and show me tiny spots on their bodies that they believed were caused by dispersants. But the facts in these types of cases are not what matters. The only thing you can really do is show compassion. And online, this is an extremely difficult task—especially when you don't know who you're dealing with.

That said, Jay Cullen is right in that we can't let scientific falsehoods go unchecked either. Because this just opens the door for more misinformation to creep into people's lives. So while engaging with individual deniers on social media is likely to be unproductive, you can still post information that is relevant to an event. Your post may not get the facts to people who don't want to hear them. But it might get the facts to someone within their community—someone whom they are more likely to trust. And if you can add a human dimension to your post—such as a friendly face or a story about your own vulnerabilities or challenges—all the better.

Notes

1 While neither of us is a medical doctor, we were both deeply ingrained in the spill and we both had a keen interest in the dispersant controversy. At that time, we were unaware of any dermatological issues linked to dispersant applications. A recent article on the long-term studies on the health of responders following dispersant application does not reveal any negative effects. See P. A. Sandifer et al., "Human Health and Socioeconomic Effects of the Deepwater Horizon Oil Spill in the Gulf of Mexico," *Oceanography* 34, no. 1 (2021): https://doi.org/10.5670/oceanog.2021.125.

2 See Thad Allen, "National Incident Commander's Report: MC252 Deepwater Horizon," US National Response Team, October 1, 2010, www.nrt.org/sites/2/files/DWH%20NIC.pdf.

3 F. Menczer and C. Hills, "The Attention Economy," *Scientific American* 323 (December 2020): 6, 54–61, https://doi.org/10.1038/scientificamerican1220-54.

4 L. Wells and A. Gowda, "A Legacy of Mistrust: African Americans and the US Healthcare System," *Proceedings of UCLA Health* 24 (2020), https://proceedings.med.ucla.edu/index.php/2020/06/12/a-legacy-of-mistrust-african-americans-and-the-us-healthcare-system.

5 K. Kricorian and K. Turner, "COVID-19 Vaccine Acceptance and Beliefs among Black and Hispanic Americans," *PLoS ONE* 16, no. 8: e0256122, https://doi.org/10.1371/journal.pone.0256122.

6 David Leonhardt, "U.S. Covid Deaths Get Even Redder," *New York Times*, November 8, 2021. The story on stagnant living standards is Leonhardt, "A Great Fight of Our Times," *New York Times*, October 11, 2016.

7 The Ad Council, "The History of Our Covid-19 Vaccine Education Initiative," www.adcouncil.org/our-impact/covid-vaccine/our-covid-19-vaccine-retrospective.

8 N. Light et al., "Knowledge Overconfidence Is Associated with Anti-consensus Views on Controversial Scientific Issues," *Science Advances* 8, no. 29 (2022): https://doi.org/10.1126/sciadv.abo0038.

9 C.-I. Lin, "Emergence of Perceptions of Smart Agriculture at a Community/Campus Farm: A Participatory Experience," *JCOM* 21, no. 2 (2022): https://doi.org/10.22323/2.21020202. On the use of narrative, see M. Dahlstrom, "The Narrative Truth about Scientific Misinformation," *PNAS* 118, no. 15 (2021): https://doi.org/10.1073/pnas.1914085117.

10 B. Fischhoff, "The Science of Science Communication," *PNAS* 110 (2013): https://doi.org/10.1073/pnas.1213273110.

11 Interview, *The Agenda*, January 30, 2018. www.youtube.com/watch?v=XhPQ2oSZDfI.

12 Mark Hume, "Canadian Researcher Targeted by Hate Campaign over Fukushima Findings," *Globe and Mail*, November 1, 2015.

8 Legal Challenges

The Erosion of the Scientific Deliberative Process

Science Out of Context

By Christopher Reddy and Richard Camilli
Boston Globe
June 3, 2012

Late last week, we reluctantly handed over more than 3,000 confidential e-mails to BP, as part of a subpoena from the oil company demanding access to them because of the *Deepwater Horizon* disaster lawsuit brought by the US government. We are accused of no crimes, nor are we party to the lawsuit. We are two scientists at an academic research institution who responded to requests for help from BP and government officials at a time of crisis.

Because there are insufficient laws and legal precedent to shield independent scientific researchers, BP was able to use the federal courts to gain access to our private information. Although the presiding judge magistrate recognized the need to protect confidential e-mails to avoid deterring future research, she granted BP's request.

It is the lack of legal protection that has us concerned.

The 2010 *Deepwater Horizon* disaster caused the death of 11 people and spilled oil at an unprecedented depth of nearly a mile under the Gulf of Mexico. That deep-sea environment was aqua incognita to the oil industry and federal responders, but a familiar neighborhood for us at Woods Hole Oceanographic Institution. BP and Coast Guard officials asked for our help to assess the disaster, and we obliged.

We responded by leading on-site operations using robotic submersibles equipped with advanced technologies that we had developed for marine science. We applied them to measure the rate of fluid release from the well and to sample fluids from within the well. We then volunteered our professional time to scrutinize this data and published two peer-reviewed studies in a respected scientific journal. We determined an average flow rate of 57,000 barrels of oil per day and calculated a total release of approximately 4.9 million barrels.

BP claimed that it needed to better understand our findings because billions of dollars in fines are potentially at stake. So we produced more than 50,000 pages of documents, raw data, reports, and algorithms used in our research—everything BP would need to analyze and confirm our findings. But BP still demanded access to

DOI: 10.4324/9781003341871-11

our private communications. Our concern is not simply invasion of privacy, but the erosion of the scientific deliberative process.

Deliberation is an integral part of the scientific method that has existed for more than 2,000 years; e-mail is the twenty-first-century medium by which these deliberations now often occur. During this process, researchers challenge each other and hone ideas. In reviewing our private documents, BP will probably find e-mail correspondence showing that during the course of our analysis, we hit dead-ends; that we remained skeptical and pushed one another to analyze data from various perspectives; that we discovered weaknesses in our methods (if only to find ways to make them stronger); or that we modified our course, especially when we received new information that provided additional insight and caused us to reexamine hypotheses and methods.

In these candid discussions among researchers, constructive criticism and devil's advocacy are welcomed. Such interchange does not cast doubt on the strengths of our conclusions; rather, it constitutes the typically unvarnished, yet rigorous, deliberative process by which scientists test and refine their conclusions to reduce uncertainty and increase accuracy. To ensure the research's quality, scientific peers conduct an independent and comprehensive review of the work before it is published.

A byproduct of the order to hand over our e-mails is that BP now has access to the intellectual property attached to the e-mails, including advanced robotic navigation tools and sub-sea surveillance technologies that have required substantial research investment by our laboratories and have great economic value to marine industries such as offshore energy production. The court provides no counterbalancing legal assistance to verify that BP or its affiliates do not infringe on our property rights. Although there is a confidentiality agreement that BP is subject to, the burden is left entirely to us, a single academic research organization, to police the use of our intellectual property by one of the largest corporations in the world.

Ultimately, this is not about BP. Our experience highlights that virtually all of scientists' deliberative communications, including e-mails and attached documents, can be subject to legal proceedings without limitation. Incomplete thoughts and half-finished documents attached to emails can be taken out of context and impugned by people who have a motive for discrediting the findings. In addition to obscuring true scientific findings, this situation casts a chill over the scientific process. In future crises, scientists may censor or avoid deliberations, and more importantly, be reluctant to volunteer valuable expertise and technology that emergency responders don't possess. Open, scientific deliberation is critical to science. It needs to be protected in a way that maintains transparency in the scientific process, but also avoids unnecessary intrusions that stifle research vital to national security and economic interests.

The *Deepwater Horizon* Subpoena

In 2011, BP faced over one hundred lawsuits from the US government, the fishing industry, and residents of the Gulf Coast states. The US Department of Justice lawsuit assigned blame for the spill to BP, with the initial estimated penalties for environmental damages totaling over US$20 billion.

According to the Clean Water Act, the formula for assessing such penalties appears straightforward: a maximum of $4,300 for each barrel of oil spilled. Thus, the big number to be determined was: how much oil actually spilled?

As recounted in Chapter 5, the Woods Hole Oceanographic Institution (WHOI) team played a critical role in determining the government's final spill estimate of 4.9 million barrels. While we were not a party to the government's lawsuit, our voluntary role in determining the flow rate wound up entangling us in the court case all the same.

So far, I've largely discussed why and how scientists should communicate with other stakeholder groups. But there are also times when it is in your interest to not communicate. You may find yourself in a situation that makes you feel uncomfortable: perhaps a journalist is pressuring you to say something you don't agree with. Or you are being asked for a statement on a subject that is not your area of expertise. Or you may feel like you are not the best person to convey your message effectively—some people simply aren't good candidates to be interviewed, and that's okay.

But the use of legal means to harass and intimidate scientists is another matter entirely. In this case, we could not say *no*. While being embroiled in a lawsuit may sound extreme, unfortunately, the harassment of scientists through subpoenas or open records requests has become increasingly common. As discussed in Chapter 3, University of Minnesota chemist Deborah Swackhamer found out that when well-financed companies or organizations feel threatened by a scientist's research, legal intimidation is one way to slow down or stop funding entirely.

Although the 1966 Freedom of Information Act was passed to increase transparency and accountability in the government, it has also been turned on its head to harass scientists at public institutions, by allowing threatened parties to demand every scrap of information about a research project, from marked-up rough drafts and handwritten notes to private telephone records. These sorts of tactics are designed to go beyond simply gaining information needed for reproducibility—they aim to put the maximum amount of psychological and financial pressure on the individuals involved.

As shown in Table 8.1, examples of industry pressure ranges from Dow Chemical and Monsanto to big tobacco, oil, and gas: no surprises there. In contrast, the targets of these subpoenas and open records requests are considerably more varied. Independent scientists, medical researchers, interns, peer reviewers, book publishers, NOAA, and even the NSF have all been targets of harassment. In one particularly egregious example, the Australian Department of Climate Change received over 750 freedom-of-information requests over a four-month period in 2011, from a single organization, the Institute of Public Affairs, with each request taking roughly thirty-nine hours to complete.[1]

For WHOI, the work that we voluntarily did on behalf of the Coast Guard and Incident Command in June 2010 became the subject of similar intense scrutiny. Our principal task at that time was to estimate the amount of oil

Table 8.1 High-profile legal intimidation cases against scientists

Party Issuing the Subpoena or Open Records Request	Scientific Institution or Individual Researchers	Reason for Subpoena or Open Records Request	Year	Outcome
Dow Chemical	University of Wisconsin	Attempt to bolster claim that scientific studies on carcinogenic properties of Agent Orange were inconclusive.	1980	The Federal Court sided with university researchers.
Philip Morris and R. J. Reynolds	Mount Sinai Medical School	Demanded data and documents from two published lung cancer studies in order to publicly discredit said studies.	1989	The Federal Court sided with the tobacco companies.
Anonymous clients of a Wall Street law firm	Deborah Swackhamer, University of Minnesota, EPA	Attempt to impede research and stop EPA funding on toxaphene pollution in Great Lakes.	1996	The largest open records request ever filed in Minnesota: "We shipped off container after container of papers; they kept coming back for more and more information."
Dow Chemical, Monsanto, Goodrich, Goodyear, Union Carbide, and fifteen others	University of California Press, Gerald Markowitz, David Rosner, peer reviewers of the publication, and the National Science Foundation	Attempt to discredit the authors, publisher, and peer reviewers of the book *Deceit and Denial: The Deadly Politics of Industrial Pollution* (University of California Press, 2002), which unveiled attempts by companies to conceal the link between their products and cancer.	2005	Week-long deposition of the two authors and five peer reviewers in court; the requested documents were eventually all submitted to the court.

Party Issuing the Subpoena or Open Records Request	Scientific Institution or Individual Researchers	Reason for Subpoena or Open Records Request	Year	Outcome
Virginia Attorney General Ken Cuccinelli, American Tradition Institute	Michael Mann, University of Virginia	Attempt to discredit climate change science.	2010	The Virginia Supreme Court rejected both the subpoenas and the open records requests.
BP	Richard Camilli, Christopher Reddy, WHOI	Attempt to discredit the methodology used to determine the flow rate of oil leaking from the wellhead.	2012	The Federal Court sided with BP; all deliberative documentation handed over. However, BP was required to pay all of WHOI's legal fees.
Science, Space, and Technology Committee Chairman Lamar Smith, US House of Representatives	National Oceanic and Atmospheric Administration (NOAA)	Attempt to discredit NOAA's report on climate change, published in *Science* on June 3, 2015.	2015	NOAA provided the first tranche of requested documentation but stopped short of turning over private emails and correspondence between scientists not involved in the study.
Science, Space, and Technology Committee Chairman Lamar Smith, US House of Representatives	Union of Concerned Scientists, seven other science and environmental organizations, and the attorneys general of New York and Massachusetts	Impede investigations—and the attorneys' own subpoenas of Exxon Mobile—into whether Exxon Mobile committed fraud and suppressed its own scientists' findings with regard to the risks of climate change.	2016	The attorneys general offices refused to comply with the subpoena, stating it was invalid.

Sources: M. Halpern, "Freedom to Bully," *Union of Concerned Scientists*, February 15, 2015: www.ucsusa.org/resources/freedom-bully; R. Camilli, A. Bowen, C. M. Reddy, J. S. Seewald, D. R. Yoerger, "When Scientific Research and Legal Practice Collide," *Science* 337, no. 6102 (2012), https://doi.org/10.1126/science.1225644; J. Johnson, "Congressional Panel Battles over Subpoenas of Scientists," *Chemical and Engineering News*, September 15, 2016; M. Henry, "House Committee Chairman and NOAA Administrator Spar on Climate Change Study," *American Institute of Physics*, May 12, 2016.

leaking from the well. This was done in two steps, both of which were headed by WHOI scientist Rich Camilli. The first step was to determine the volume of oil and gas leaking from the well. The second step was to determine the ratio of oil to gas at the wellhead, which allowed us to more accurately determine the overall oil flow rate. This number contributed to the government's final spill estimate of 4.9 million barrels.

BP, of course, wanted to reduce their financial liability in the face of the Department of Justice (DOJ) lawsuit, so their legal team got to work, trying to figure out avenues through which they could achieve this. And one of their tactics was to call into question our calculated flow rate. On December 8, 2011, Rich and I received a letter from the federal courts. BP was subpoenaing us to produce everything we had that was related to the study. And when I say everything, I mean *everything*.

In some respects, we were lucky, because WHOI stood up for us. Our institution immediately issued an objection to the comprehensive scope of the subpoena. They told us they would pay for all of our legal fees, win or lose. Other researchers who were also subpoenaed were not so fortunate. Their institutions simply said, "Just give them whatever they ask for; we're not going to pay for you to hire a lawyer."

In our case, we didn't object to turning over the raw data and related documentation, but we thought it was highly unethical to have to give up our deliberative discussions, particularly because of the precedent it might set. After several months of back and forth, including with the US Coast Guard, with whom we had signed a confidentiality clause before undertaking our research, we eventually submitted 50,000 pages of documentation and 34GB of data. This was more than enough information for BP to reproduce the peer-reviewed findings we had published in the *Proceedings of the National Academy of Sciences* earlier that year.

But still, BP wanted more. Specifically, they wanted the rough drafts of the two papers we had published in *PNAS*, with all of our comments and tracked changes. They wanted to see if we were making revisions that were in bad faith, or which exhibited bias, or which were indicative of poor science. Both Rich and I put up a fight—we felt that giving them all of our data, plus two peer-reviewed published papers, should have been enough. But as our lawyer kept reminding us, they were looking for whatever daylight they could find.

BP eventually asked federal magistrate Sally Shushan for assistance in enforcing the subpoena, and in April 2012, she issued an order stating that while she recognized WHOI's arguments in filing an objection, we still had to comply with the original request and turn over all of our deliberative communications.

It is impossible to convey just how much pressure we were subjected to during this lawsuit—in which, it bears reminding, we were not even a party. Over the course of six months, I wound up losing seventy-five pounds. My marriage nearly fell apart. I had a nervous breakdown, and had to take repeated mental health leaves of absences. And I was in three separate car accidents, suffering a concussion each time. In the end, the process broke me.

My oldest friends know that there was a pre-*Deepwater Horizon* Chris, and now there is a post-*Deepwater Horizon* Chris.

If we had known what the consequences would be for assisting the Coast Guard with the flow-rate calculations, would we have done it? This question was the whole reason we fought the scope of BP's subpoena. As Susan Avery and Laurence Madin, then president and director of research for Woods Hole, respectively, wrote in an editorial,

> pulling academics and researchers into litigation they are not a party to will have a chilling effect on how science is conducted. The essence of the scientific process is rigorous deliberation in which scientists examine, question, test, reject, and modify ideas as they work toward a verifiable conclusion. Without adequate legal protection, researchers and their institutions may reasonably fear that their deliberative process can be attacked and their intellectual property exposed, or that they will become entrained in litigation to which they are not parties and where they are unlikely to derive any benefit. As a consequence, scientists may feel forced to curtail, censor or avoid the normal deliberative process. In future emergencies, particularly those that might give rise to litigation, researchers may be more reluctant to volunteer expertise and technology.[2]

Box 8.1 Dealing with Personal Harassment

Unfortunately, harassment of scientists can go well beyond legal means. Online attacks, threats of physical or sexual violence, and assaults on credibility have become increasingly common in the twenty-first century. In some cases, activists may release a scientist's personal information—including home address, email, and telephone number—and encourage supporters to badger the person directly.

But with social media and email, it is usually much easier to simply troll researchers online. While COVID-19 and climate change are the two of the biggest hot-button issues, harassment can arise from any number of topics, ranging from lab animals and vaccinations to biotech and genetically modified organisms (GMOs).

In a 2021 survey conducted by *Nature*, over three hundred scientists reported harassment following a media appearance regarding COVID-19. Of the responders, "more than two-thirds reported negative experiences as a result of their media appearances or their social media comments, and 22% had received threats of physical or sexual violence." An additional 15% had received death threats, while 30% said their reputations had been damaged, and 42% reported emotional or psychological distress following the abuse.[3]

Regardless of the type of harassment you might face, how you choose to respond is entirely up to you. Some people avoid social media entirely, as they feel that having to deal with the continual stream of aggression and negativity is just not worth it. Others, in contrast, see social media as a tool that allows them to take ownership of their own narrative. Platforms like Twitter and TikTok can be useful tools for explaining what they do and why they do it. If an organization has released a lot of negative stories about you or your research, it can also be a good way to achieve a more balanced presence online.

As far as email harassment goes, most researchers ignore outright offensive or personal attacks and incoherent ramblings. But some do take the time to respond to their critics, addressing false claims and sending simple fact sheets on the subject matter to their detractors. Just remember: responding to critics takes time and energy. Make sure you have both in ample supply before you engage.

As Johanna Haigh, an atmospheric physicist from the Imperial College of London, said:

> I worry about younger scientists who can find themselves targets for attacks they are unprepared to handle. My advice is simple: play it straight. Don't rise to the bait. Explain politely what you understand and what perhaps they have misunderstood. If they are offensive, do not respond.[4]

The Legal Outcome

After the subpoena, the concussions, the nervous breakdown, and accumulated stress at home, most of my friends said to me, "See what BP did to you? This is the price you paid for trying to help. They ruined your life." But I look at it in a different way. It is both easy and tempting to demonize BP—and all of industry for that matter. But to do so misses the point. Making a villain out of someone may help you process and rationalize your suffering, but it doesn't actually solve the problem.

The issue at stake here is not that a multinational corporation was trying to reduce their financial liability in the courts—it is much more important than that. It is that, despite decades of legal harassment of scientists, there are still no legal protections for the scientific deliberative process. For scientists, there is nothing that resembles physician–patient or attorney–client confidentiality. But there should be.

In 1954, during the height of McCarthyism, New Hampshire's state attorney general, who was investigating individuals suspected of belonging to the Communist Party, held the economist Paul Sweezy in contempt of court and sentenced him to a jail term for refusing to divulge the contents of the lectures he gave at the University of New Hampshire.

The case eventually moved up to the Supreme Court, and in 1957 the justices ruled in favor of Sweezy, establishing an important precedent in the protection of academic and intellectual freedoms. In the *Sweezy v. New Hampshire* decision, Chief Justice Earl Warren and three others wrote: "Freedom to reason and freedom for disputation on the basis of observation and experiment are the necessary conditions for the advancement of scientific knowledge."[5]

And yet, in spite of this landmark case, the threat of litigation hampering the scientific process is stronger than ever. As more and more special interest groups, corporations, activists, and even elected officials file open records requests for scientific research in an attempt to sabotage the research process, increasing amounts of time, money, and staff have to be dedicated simply to replying to these legal challenges.

One of the most well-known examples is that of the former congressman Lamar Smith (Texas), who, following his appointment as the chairman of the Committee on Science, Space, and Technology in the US House of Representatives in 2012, proceeded to file over twenty-five subpoenas against climate change scientists, in addition to attempting to undermine peer review at the NSF, limit the EPA's ability to regulate polluters, and cut back NASA's operating budget.[6] Congressman Smith didn't even have to go through his committee to file the subpoenas—he was able to issue them at his own behest. So, here we have a Congressional committee designed to support scientific research and development that, it can be argued, abused its power in order to do the exact opposite. Prior to former congressman Smith's appointment as the Committee chair, the number of subpoenas the Committee had issued since 1959 was a grand total of one.

Box 8.2 The *E* in Email Stands for *Evidence*

In today's world, it's impossible not to have a record: emails, tweets, posts, texts, phone messages ... Words you never intended to go public, or that could be easily misconstrued when taken out of context ("I could just kill him for doing that!"), can always come back to bite you—whether in the courtroom or somewhere else. Thus, a reminder: think twice whenever you share your thoughts electronically. Some people even go so far as to tell their colleagues, "Don't send an email, and don't leave me a message. Let's talk about it in person." This is a good time to remember, email isn't a great communication tool anyway: there's too much room for misinterpretation, ambiguity, and hurt feelings.

At the same time, it's also worth considering how email can work in your favor. Sometimes, you want there to be a record. You want to be able to prove that you said a certain thing, at a certain time, to a certain person. In that case, email is great. Whenever you're drafting an important piece of writing, make sure you do it as a two- or three-day process. Don't write it up, reread it once, and then fire it off. Give yourself time to change your mindset, to read what you have written with fresh eyes. Maybe you'll spot a mistake. Or maybe you'll realize that your tone was a little off.

One final piece of advice about email: if you are angry or upset about something, by all means write down your thoughts and feelings in an email. Let it out. But never, ever click send. At least, not until a good amount of time has passed and you've had the opportunity to reflect more on the consequences. Too often, emotionally driven missives can turn into much larger problems—and the record of what you said will be there forever.

So what is the solution? It is to provide improved legal protection for the deliberative process of science. Because without such protection, scientific research will surely suffer. In our case, we ultimately felt like we achieved something of a moral victory, as the magistrate validated our argument and made BP pay for both our legal fees and all the time we spent gathering the requested documentation. But looking back at everything Rich and I went through, the question of "Was it worth it?" is harder to answer.

With a decade of reflection behind me, I think that if I were thrust into a similar crisis in the future, I would make the same decisions all over again. I would do the science on behalf of the government, and I would write and publish the results in a peer-reviewed journal. The only thing I would change is how I dealt with the legal battle. In hindsight, the outcome of fighting the subpoena was *not* worth it. The personal cost was too high, and we were even excoriated by other academics, who thought we were grandstanding. "Chris, I respect you as a person and a scientist," one of my colleagues told me, "but you don't have a leg to stand on."

In 2015, the federal courts finally issued a verdict for the *Deepwater Horizon* trial. Judge Carl J. Barbier ruled that BP was "grossly negligent" and was to be

E in email stands for evidence.

Figure 8.1 Scientists are trained to keep records. Emails, tweets, and text messages capture your words and thoughts. Take advantage of this superior recording system but be mindful of potential negative outcomes.

held liable for having spilled a net total of 3.19 million barrels of oil into the Gulf, after deducting the amount of oil removed through their collection efforts.

But, as Judge Barbier wrote, the evidence was "voluminous, dense, highly technical, and conflicting. ... There is no way to know with precision how much oil discharged."[7] In the end, he simply split the estimates roughly down the middle: the government claimed BP had spilled 4.19 million barrels (after collection), while BP claimed they spilled 2.45 million barrels (also after collection). And thus, with a verdict of 3.19 million barrels spilled and a record $18.7 billion in fines due, the legal end to the worst environmental disaster in US history finally came to a close.

Notes

1   M. Halpern, "Freedom to Bully," *Union of Concerned Scientists*, February 15, 2015: www.ucsusa.org/resources/freedom-bully.
2   S. Avery and L. Madin, "The Need to Protect the Scientific Deliberative Process," Woods Hole Oceanographic Institution (2012), www.whoi.edu/website/president-dir ector-emeritus-susan-avery/statement/scientific-deliberative-process.
3   B. Nogrady, "'I Hope You Die': How the COVID Pandemic Unleashed Attacks on Scientists," *Nature* 598 (2021): 250–53, https://doi.org/10.1038/d41586-021-02741-x.
4   V. Gewin, "Real-Life Stories of Online Harassment—And How Scientists Got through It," *Nature* 562 (2018): 449–50, https://doi.org/10.1038/d41586-018-07046-0.
5   S. Avery and L. Madin, "The Need to Protect the Scientific Deliberative Process," Woods Hole Oceanographic Institution (2012), www.whoi.edu/website/president-dir ector-emeritus-susan-avery/statement/scientific-deliberative-process.
6   J. Mervis and W. Cornwall, "Lamar Smith, the Departing Head of the House Science Panel, Will Leave a Controversial and Complicated Legacy," *Science*, November 5, 2017, https://doi.org/10.1126/science.aar4141.
7   John Schwartz, "Judge's Ruling on Gulf Oil Spill Lowers Ceiling on the Fine BP Is Facing," *New York Times*, January 15, 2015.

Part III

Lessons Learned

9   Starting a Conversation

How to Talk about Science

One of the low points in my career came on a Sunday afternoon in the summer of 2010. I had taken the day off to spend some time with my wife at the beach, but I was unable to relax, because that evening I had an interview scheduled with *Good Morning America*. This was the most high-profile interview of my life, and even though my wife Bryce kept reminding me "It's only two minutes," I knew that two minutes of television footage was likely going to equate to two hours of filming.

At the time, I was both physically and mentally exhausted—it was right in the middle of *Deepwater Horizon*—and the last thing I wanted to do was go back to my lab. But the chance to appear on *Good Morning America*? For weeks, all I had been seeing in the news was one catastrophic speculation after another. Speculations that were hurting the people who lived along the Gulf Coast, people who were still facing enormous uncertainty and stress. *Good Morning America* has a huge reach, and thus I thought, rather naively, that it would be a chance for me to shift the course of the narrative. And what's more, I knew that if I didn't do the interview, they would find someone else who would. What if that someone else did a poor job?

The excitement of appearing on a national television program can be overwhelming, and my nerves, combined with my generally fatigued state, helped to set the stage for the disaster to come. But nerves aside, by this point in my career, I should have known better. I should have known that television crews will often ask you to do things that are not realistic, or have you take off your safety glasses or position you in front of a background that has nothing to do with the topic.

This alone was causing frustration that I should have been able to manage. When someone asked me to improperly mix some chemicals in a volumetric flask for cutaway footage, all I could think was, "I hope none of my colleagues see this, because I am going to look like a complete idiot!"

Another source of frustration was the journalist himself, who was conducting the interview remotely. This was yet another shortcut I had taken—normally, whenever I'm interviewed, I always check out the reporter first and see what sort of stories they cover. But because it was my day off, I didn't bother to do any research. I had no idea who he was. The only thing I could

DOI: 10.4324/9781003341871-13

recall from WHOI's media relations manager, Stephanie Murphy, was that he was a handsome television type who had been covering *Deepwater Horizon*.

When it came time for the interview, I realized too late that our interpretations of the crisis were fundamentally at odds. He was expecting me to validate his fear-based narrative, while I was hoping to break down some of the more common misperceptions about the spill. The interview sputtered along, with him trying to push me for a comment on the National Oceanic and Atmospheric Administration (NOAA), and me refusing to play ball.[1] As Stephanie later recalled,

> The classic technique when you don't want to answer a difficult media question is to block and bridge. You deflect the original question and move on to what you really want to talk about. And that night, I kept waiting for you to bridge and you never did. Instead, you just kept repeating, "I'm not going to answer that," over and over again.[2]

Things continued to escalate until I finally lost it. I am not proud of what I said, and I like to think if I hadn't been so tired, if I hadn't already spent two hours in front of a television crew when I should have been home with my wife, I would have acted with considerably more civility.

But instead, I blurted out, "Listen, I'm not going to tell you how to blow dry your hair if you don't tell me how to do science!"

Well, that was the end of that interview. The television crew was out of there so fast I couldn't believe it had taken them two hours just to get some footage of me mixing chemicals in a volumetric flask.

I knew immediately that I had blown it. I had insulted the entire crew of *Good Morning America* and missed a tremendous opportunity. Rather than making a positive impact, I had made no impact at all—except to reinforce the exact stereotype that drives me crazy: scientists are elitist assholes. Stephanie was not pleased with me either, telling me in the aftermath, "I wish you could have given them something they could use."

The lesson here is twofold: The first is that if you're going into a media interview, you should always be prepared, because that will allow you to figure out what your message is and how you are going to get there. Reporters are not always going to lob you softballs, and you should be ready for tough questions. But the more important lesson is that there's always a middle ground. Even though we had competing goals, I should have been able to figure out a way to give him what he needed while also managing to provide real content as a scientist. Because, in the grand scheme of things, the fact that he pulled the plug meant that science was the biggest loser of all.

So, What Do You Do?

Whenever I have any kind of discussion about science, I want it to end on a positive note. However, as this example illustrates, achieving this goal is not always as easy as it seems. Scientists, already misunderstood in the past, have increasingly become the targets of derision, mockery, and even hatred. As our society has become more complex and dependent on technology that people don't understand, with that incomprehension has come fear. And with that fear has come a resistance to change. You might say that science is Pandora's box, and the more people hear about topics like cloning, artificial intelligence, and bioengineering, the more people feel that the world would be better off if that box had remained closed from the beginning.

Regardless of COVID-19's origins, many people around the world believe that the virus was created in a lab. They will never be convinced otherwise. *Science*, those people say, *is at fault. These people are meddling with things that should not be meddled with.* Forget about penicillin, the smallpox vaccine, and heart transplants. Electricity, pasteurization, and modern transportation. It only takes a major crisis, combined with a shred of uncertainty, and we are back at the bottom of the Sisyphean hill. As with so many other examples, scientists are caught in the web of complexity that engulfs all of us.

When faced with someone who is aggressively spouting off misinformation or challenging your credentials, it is all too tempting to return tit for tat. But where does this lead? Does this help people understand science? It certainly won't make them accept it—conflict only leads to more conflict. So how do we reach the ever-elusive goal of positive impact?

It starts with conversation.

A conversation is not a one-sided lecture. A conversation requires at least two people, and both of those people have to participate. Just as I don't want to get into an argument when talking with someone, nor do I want them to stifle a yawn, excuse themselves, and run off for a drink after only two minutes. My ultimate goal is for the other person, or people, to remember our conversation long after it ended. For it to make it to the dinner table later that night: "You'll never guess what I learned today ..."

This may sound like an egotistical attitude. But I'm not suggesting we talk about ourselves more. I'm suggesting we should talk about *science* more. It's not every day that people meet a scientist, so why not make it count?

No matter your personality type—whether you're introverted or extroverted, concrete or abstract, goofy or earnest—as scientists, there are certain obstacles that we must all overcome. One of those is a tendency to think more about the details than the big picture. When faced with the common icebreaker—"So what do you do?"—how do you answer? Do you list every line on your three-page résumé? Do you launch into a ten-minute summary of what sunspots are, the fluctuating temperatures on the sun's surface, and how they affect the Earth?

In the same way that trying to figure out what makes other people excited is a good conversation skill, remembering what makes *you* excited is equally important. Why do you wake up in the morning? What are you trying to solve? Remember, enthusiasm is contagious.

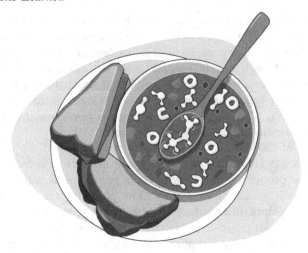

Start in your element.

Figure 9.1 Communicating science is hard. Build your skills in your element among friends and family. Countless studies have shown that conversations over food create relaxed and open exchanges.

Too often I see scientists talk about the technology they use instead of the research questions they have. While we don't have to talk about our work in order to have an engaging conversation about science, if someone does ask you what you do, it's helpful if you can figure out how to relate your world to theirs.

In my case, I don't ever use the words *gas chromatography* or *chemistry*. Who cares about that? Instead, I say, "I study how nature responds to the uninvited," or, "I study what happens when pollutants are introduced into the environment." People do care about pollution, and this simple line is usually enough to launch a conversation. Maybe we'll get to talking about plastic bags—a topic I find absolutely fascinating, and which many people feel quite strongly about—or pesticides, or fertilizers, or household cleaning supplies ... I've found there's no shortage of chemistry-related subjects that people love to talk about—just so long as you don't mention the word *chemistry*. And what about that glass of wine they're holding? Why is theirs white, and why is mine red?

No matter how the conversation progresses, remember, "What do you do?" is no more than a launchpad. People probably don't want to hear about your research in great detail—at least, not right away—but they will respond to your enthusiasm. And if you can figure out how to make them feel special, and how science ties into their interests, then you're off to a good start.

Box 9.1 Chemist, Scientist, Or Oceanographer? Lessons in Perception

How much do titles and perceptions matter? Consider this: the way you introduce yourself to someone has a huge impact. Over the course of my life, I have tried a number of different responses to the question, "So what do you do?" and found that the reactions I get vary significantly depending on how I frame my answer.

If I tell someone I'm a chemist—probably the most accurate description of my day-to-day world—more often than not their nose will wrinkle up, as if they can smell the volatile compounds wafting in their direction. And then they say, "I don't know anything about chemistry. The last time I did chemistry was in eleventh grade, when I almost burned down my high school." Now, I find it hard to believe that almost every stranger I've ever met has almost burned down a school, but the message is clear: chemistry is not cool.

In contrast, if I take a more general approach—by saying I'm a scientist, for example—it can be hard for people to relate. The term *scientist* is just too broad; people don't know what it means, other than conjuring up a dorky, last-pick-in-gym-class personality. Whereas *chemist* elicits a negative reaction, *scientist* often elicits no reaction. There is just no emotional connection.

But if I tell someone I'm an oceanographer, then their eyes light up immediately. "Oh, do you scuba dive? Do you work with whales? Are you, like, Jacques Cousteau?" Of course, I don't know how to scuba dive, but that doesn't matter. The point is, how people react to what you say depends on their personal connection to the words you use. Oceanographers are cool because people think they are out there saving the world and swimming with whales and tropical fish. Chemists are bad, because high school chemistry can be boring and disconnected from what people care about. Instead of triggering curiosity, it killed curiosity. *Okay, we are now going to do the same lab that every kid who has come through this school for the past two decades has done, and you are going to get the same result as all of them, and if you don't, then you must have done it wrong and you've failed.*

No way! That's not what science is about. But it does illustrate the perceptions we have to overcome when talking about science: it is boring and uncool. Keep this in mind when you're talking to someone and see if you can figure out what about science makes their eyes light up. It will make all the difference.

Learning to Share

> Chris, that sounds like good advice, but I'm really shy. I find it painful to engage in small talk with strangers.

The truth of the matter is … it doesn't come easy to me, either! For years, I was extremely self-conscious. I hated my Rhode Island accent. People didn't hesitate to make fun of me straight to my face because of the way I talked. I also have trouble making eye contact. I'll look off to the side, or down at my shoes, or into the distance—my gaze might be anywhere but on the face of the person I'm talking to. Unfortunately, there's no solution to this but to practice. And if you want practice, then you should start in a safe space, a place where there is little risk or pressure.

And where might that be? One place to consider is your family. I know, I know ... not everyone gets along with their families. We'll talk about that too. But families can be a good place to start because you'll likely have a diversity of opinions and life experiences—as well as your own shared connections. Even if you disagree about certain things, you do have a special bond that ties you together. And you are often stuck together for at least one big meal over the course of the year.

The flip side of that coin is that sometimes *we're* the ones who make things difficult for our families. We're the weird cousin who spent their childhood tinkering in the garage or trying to raise an eclipse of gypsy moths in a Folger's coffee can. We're the ones who drifted away during college, and never managed to reestablish a connection. We're the ones who feel embarrassed or defensive when our parents try to learn more about our lives.

Let's look at an example. One of my graduate students recently had her first ever paper accepted for publication. When she told me, I congratulated her and then said: "Can you do me a favor? Forward the email the editor sent you to your parents. Tell them that getting this paper published is a big deal, and make sure you explain why it's a big deal."

After I suggested this, her face flushed and she shook her head. "Oh, I'm not going to send this to my parents."

"No," I said, "you have to send it to your parents, because it means something to them. Your parents love you, and they want to be able to share your good news."

While she agreed with me in principle, she continued to resist the idea, even going so far as to say, "I just don't feel comfortable bragging about it, even to them."

I understood her sentiments exactly—and that was why I was so adamant that she reach out. Because I was the exact same way for many years. For the longest time, whenever I went somewhere with my dad, he would always introduce me as, "My son, the world-renowned scientist." Of course, whenever he said this, my stomach would churn with embarrassment. You can probably chalk some of that up to a parent's uncanny ability to mortify their children, even in adulthood. But our scientific training also plays a role—we have been conditioned by the pervading culture to not grandstand, ever.

After complaining about my father's line one too many times, Bryce, in exasperation, finally said, "This isn't about you. Do you realize how much joy your dad gets by saying that? His introducing you that way is probably the best part of his day." That was my wake-up call. And so, ever since, I started sending regular updates to my parents. If I'm quoted in a newspaper or magazine article, or interviewed on the radio, I'll send them a link, because I know it will make them happy.

One reason why sharing your successes with your parents is so important is because there is often no risk involved. Your colleagues are not going to think you're a showboat if you tell Mom and Dad good news. Now, your parents might not understand the import of what you're telling them, or they might even disagree with you, but it doesn't matter. When it comes to outreach, your family can be a great place to start. And the reward is that you will make them proud.

Box 9.2 Family Time

Part of being a good communicator is knowing when the right time to speak is. Just because *you* are passionate about science doesn't mean that the people you love are going to feel the same way.

For example, Bryce and I used to enjoy watching the crime show *CSI* together. But there was one problem: my running commentary on the feasibility of whatever was happening would drive her crazy. I would say things like, "They would never get DNA results that fast!" Or, "That chemical is colorless, why did they make it blue?" Or, "That's not the right word. They don't mean *precise*, they mean *accurate*."

This would inevitably be followed by, "Honey, shush. I'm trying to watch the show!"

I tried. I really did. But eventually, we had to work out a compromise. I was allowed to have one "They can't do that!" per episode, but the rest of the time, I had to keep my mouth shut. Your family probably doesn't want to be in science class when you're just hanging out together, and that's okay. It just means that when you do choose to explain something, they're more likely to listen.

It's Okay to Walk Away

> Chris, I have gotten into horrible arguments with my family about science. This isn't a low-risk situation at all. It is an extremely emotional topic and threatens the core of our relationship.

This, too, is a place that many of us have been. COVID-19, climate change, renewable energy, vaccinations in general—heck, even the age and shape of the Earth—can all be explosive issues. Debating them can cause you real pain, and only you know the best way to proceed. But I do want to remind you, it's always okay to walk away. Walking away doesn't mean giving up. It means acknowledging that your relationship with someone who disagrees with you is more important than the disagreement itself. And sometimes, that can make the difference in the long run.

During my teens and early twenties, from middle school through college, I was a wrestler. I was the only science major on my wrestling team at Rhode

Island College, and I went on to be named a Scholar All-American. Wrestling was something that brought me a lot of joy throughout these years, and I came to think of my teammates and coaches as a second family. Even though we all wound up taking different paths in life, I continue to remain close to most of those guys.[3]

So when I was invited to a reunion banquet for my high school team a few years ago, I was super excited to see everyone again. We were all seated together at the same table, alongside our coach, who, by this point, I had probably known for at least twenty-five years.

As the evening drew on, most of the guys had had a few beers, and things began to get a bit rowdy. It was at this point that someone asked, "Hey Chris. Is climate change for real?"

And so I replied, "Oh yeah, absolutely." Everyone knew that I had a PhD in chemistry, but for the most part we never talked about science. So when the topic came up, I was, in my usual way, excited to start a conversation. My coach, however, had other ideas. Instead of engaging in a real discussion, he started egging on the rest of the table by mocking me and turning it into a big joke.

Before I knew what was happening, he was telling everyone that climate change was just a big hoax. And he knew it was a big hoax because his daughter had just done a project for her middle school science fair. And in her project she proved it wasn't real, because the temperature on Mars was rising too, and there aren't any people on Mars. And what did I have to say to that?

At that point, I just stopped talking. Although I was sober, everyone else had been drinking, and the tone had become aggressive. Unlike the interview with *Good Morning America*, I was able to exercise more self-restraint that night. I knew that I had to walk away, because if I opened my mouth again, I wouldn't have been able to stop myself. I would have lashed out with a hammer, and whatever I hit would have been broken beyond recognition.

And that was the thing: I was among a group of friends I had known my entire life. Outside of that moment, my coach had always been good to me. I was fond of everyone at that table, and I didn't want to ruin our relationship. And so I simply walked away and, aside from my bruised feelings, the night ended without consequence.

You might not agree with the way I handled the situation. You might think, "Maybe you managed to maintain your friendships, but that guy pushed his uninformed agenda to the rest of the table, while you said nothing."

But here's the thing: the people at that table who might have been in the middle on climate change? Who were undecided? I don't think the wrestling coach convinced them of anything at all. If anything, he pushed them in the other direction. They might have laughed in the moment, but not the next day. They all knew what I did for a living. I didn't need to remind them or rub their noses in it. And I did make clear what my opinion on the matter was.

Whatever the case was, I thought that was the end of the story. I let go of my hurt feelings and moved on with my life. But, it turned out there was one final chapter to come.

At some point in 2019, that same coach wrote me an email. He wanted my advice on harmful algal blooms and red tides. And so I replied with my thoughts on the issue. After that, he wrote back and said, "You know, I just want to apologize for the way I treated you the last time we saw each other. I was wrong and I shouldn't have said those things."

What made me happy was that, at the end of the day, when my coach had a genuine question about oceanography, he wasn't afraid to approach me and ask for my opinion. In the long run, he came around. Not everyone will, but if you leave the door open, sometimes people do walk through. They just have to do it in their own time.

Notes

1 Although I don't remember the specifics of the question, he was likely asking about the controversial NOAA pie chart, discussed in detail in Chapter 6.
2 Personal communication.
3 In fact, many of my teammates used to brag about their "smart" friend. I was even enlisted once to explain how nuclear reactors work to a woman one of my teammates was trying to impress. It must have worked, because they got married.

10 Interdisciplinary Teamwork

Plastic pollution is a complex challenge, requiring a variety of different approaches across disciplines. On any given day, there are approximately six thousand container ships, piled high with all the world's stuff, circling the globe. Factor in the threats that such ships face—inclement weather, collisions, and piracy—alongside the illegal disposal of unwanted goods in international waters and all the related carbon emissions, and it's no surprise that the maritime industry has become inextricably tied to the problem of ocean pollution.[1] During COVID-19, exhausted crews, delayed ship maintenance, and backed-up ports all helped to augment the risks.

One of the more recent accidents involving a container ship took place eleven miles off the west coast of Sri Lanka, near the capital city of Colombo. On May 20, 2021, the *X-Press Pearl*, which was waiting to offload a container of leaking nitric acid, caught fire en route from the United Arab Emirates to Malaysia, and had to make an emergency stop in Sri Lanka's highly trafficked coastal waters. In the days that followed, the Sri Lankan Navy and Coast Guard, together with the Indian Coast Guard, fought to bring the chemical fire under control, evacuate the crew, contain the leaking nitric acid, and keep the ship afloat. Although the fire was extinguished on June 1, there was nothing they could do to stop the ship from beginning to sink the next day; it finally became completely submerged on June 17.[2]

On board were 1,486 containers, containing building materials, foodstuffs, hazardous chemicals (including the already leaking twenty-five metric tons of nitric acid), heavy fuel oil (more than three hundred metric tons), and a host of other goods. The fuel oil and hazardous chemicals were the primary concern, but what went on to have the biggest environmental impact was something that passed unnoticed in the initial risk-assessment phase: nurdles—seventy-eight metric tons' worth. These tiny spherical pieces of plastic, roughly 3/16 of an inch in diameter, are one of the building blocks of modern consumerism. Melted down and poured into molds, they serve as the raw material for plastic products everywhere.

This was hardly the world's first nurdle spill. Spilled nurdles began to enter the marine ecosystem in the 1970s, and each year, another 230,000 metric tons of nurdles find their way into the planet's waters—it's all part of the chronic problem of plastic pollution in the ocean. Nurdles and other types of

DOI: 10.4324/9781003341871-14

microplastics are now present in such great quantities that they have entered the marine food web.[3] While pristine nurdles are not toxic in and of themselves, during a spill, their buoyancy, small size, and sheer numbers mean they are hard to contain. They can present choking problems for fish and other marine life, and as they age and weather, they act like sponges for toxins such as PCBs and DDT.

For an idea of what happens when a container full of nurdles falls off a ship, imagine you are scuba diving near a reef. But instead of translucent waters, schools of fish, brilliantly colored coral, and the occasional turtle, what you see instead is a blizzard of white dots—a nurdle spill is the equivalent of an underwater snowstorm. That's how many millions of tiny pieces of plastic were floating around during this spill. Sooner or later, these nurdles get washed up on shores—sometimes thousands of miles away from the original accident—in layers so thick that they can smother beach habitats.

In this particular case, the nurdle spill was further complicated by the fire on board the *X-Press Pearl*—both pristine and burnt nurdles had entered the environment, and the two did not look or behave in the same way.

The Scientific Contribution to the Spill

Right after the fire on the *X-Press Pearl* started, I received an email from one of my closest friends, Lihini Aluwihare, who is a geochemist at the Scripps Institution of Oceanography. Lihini briefed me on the crisis: nitric acid was already leaking, and several other chemicals on board—urea fertilizer, sulfuric acid, ethanol, and sodium hydroxide, in addition to the ship's fuel—were in danger of spilling right off Sri Lanka's coast. As we were both trained in organic geochemistry, she reached out to me for a second opinion, particularly as I had decades of experience with ocean pollutants.

At the same time, she also introduced me to Asha de Vos, an internationally recognized marine biologist who founded the Sri Lankan conservation NGO, Oceanswell. As a well-known ocean expert and prominent local figure in Sri Lanka, Asha had become the de facto scientific spokesperson on the ground, and was immediately inundated with question after question about the pollutants and possible long-term environmental effects.

Together, the three of us went through a crisis science checklist that would have made Steve Lehmann proud: We discussed which pollutants might pose a problem and which were not threats. We talked about which organization had been hired to do the cleanup work. We outlined the best steps to take during an oil and chemical spill. We talked about avoiding doomsday speculations, which drive fear and anxiety among the affected public. I also emailed Asha previous op-eds I had written on the role of science in a crisis during *Deepwater Horizon*, COVID-19, and the Mauritius oil spill in 2020. I could have stopped there, and it would have been enough. But I sensed there was more that we, as scientists, could contribute to the crisis, and so I continued to offer my assistance as part of the team.

Asha, who has long been devoted to scientific outreach, began fielding popular questions on social media—from acid rain to nurdles—and then posting responses from Lihini, myself, and other experts from around the world. However, because of the sheer volume of interest from the affected public, I suggested that the Woods Hole Oceanographic Institution (WHOI) communications department help her put together an FAQs page so that she could make better use of her time. We took a subset of the questions people were asking and had a half-dozen of our scientists answer them. One of WHOI's top science writers, Ken Kostel, finalized the text. Our graphic designer then created a webpage that answered the most popular questions, and provided links to further resources. Asha's name and a link to Oceanswell were placed at the top of the document.[4] And this collaborative effort was our first step in the *X-Press Pearl* communications strategy.

Soon after the nurdles began washing up on shore, it became clear that the plastic pollution was going to be the most impactful part of the crisis. And when we learned that some of these nurdles had been burnt, I immediately suggested to Asha that she collect samples and send them to Woods Hole to be analyzed; because some of these burnt nurdles didn't even resemble plastic pellets—they looked like little pieces of seagrass. The responders and cleanup crews might not even think to identify them as pollutants if they weren't informed.

Fact Sheets

After we received the samples and performed an analysis in the lab, I knew we had to communicate that there was a continuum of nurdles littering the beaches in Sri Lanka. They were different colors, different sizes, and different shapes. And the best way to present this information, I thought, would be through a fact sheet.

Publishing fact sheets in the midst of an environmental crisis may not sound like a game changer, especially during the all-hands-on-deck stage of a disaster response. But here's the thing. If it's done correctly, it works. It effectively communicates need-to-know knowledge that can be shared across stakeholder groups and will reach large numbers of people quickly. But while I was convinced of the importance of the fact sheets, not all of my colleagues thought of them as a priority. "This is obvious," was a common refrain. "There's nothing groundbreaking here. Size, color, and shape are not novel. We don't need to communicate this information." In the academic world, this may be the case: you are probably not going to get tenure with a ruler, an analytical balance, and an iPhone camera. But we weren't working with other academics.

Thus, it bears remembering that what is often obvious to scientists is *not* obvious to everyone else. And even if people might suspect something to be true, without any supporting data, they have no way of knowing whether or not they're correct. Furthermore, the idea that scientists are publishing information at the onset of a crisis can be comforting. It helps lessen both anxiety and misguided speculations, and shows that there are professionals who are actively working toward a solution.

Scientists have been trained to think that what you choose to publish has to be cutting edge or change the field in some way. But sometimes, the information that will have an impact is quite different from what you would submit to a peer-reviewed journal. Responders and the general public—and even other scientists—do not always need highly specialized knowledge. But they do need usable knowledge. They do need good data.

The main focus of our fact sheet was thus on the nurdles that had been burnt in the fire.[5] In the lab, we recognized that the burnt nurdles were going to behave differently from the unburnt nurdles in terms of toxic impacts and identification, and we wanted responders to know this as soon as possible. The people who were trying to clean up beaches and coastal waters needed to realize that they couldn't deal with the plastic pollution using a single, unified response. Burnt plastic and unburnt plastic are different entities. We weren't telling anyone what to do. The fact sheet just said, "Hey, this is what we learned. This is why it's important. This is why this information might matter to you."

Not long after we began distributing this fact sheet, Asha happened to run into a responder by the name of Conor Bolas on the beach near her home.[6] Conor was working for ITOPF, the international organization that was handling the cleanup. When she told me about the encounter, I was unable to suppress my excitement. "Did you get his card?" I asked. "That's exactly the guy who needs to have this fact sheet." So thanks to Asha's fortuitous meeting, we were able to establish a backchannel relationship with the responders at the onset of the cleanup, and they were able to reach out to us whenever they needed an analysis from our lab.

It was the type of mutually beneficial relationship that should be in place for any environmental crisis, but so often it just doesn't happen. And while Asha and Conor met by chance, part of the reason it worked out is because Conor knew who I was. So, in that sense, our successful partnership hinged on a preexisting relationship.

Eventually, we heard secondhand that the fact sheet was a big success: The Sri Lankan government used it in their briefings. The responders took the information and integrated it into their decision tree. And even the media, from news agencies in Hong Kong to the United States, found it to be a valuable resource.

Box 10.1 Crowd-Sourced Research: Turning Citizens into Scientists during a Crisis

As with any environmental crisis, one of the things that happens in the aftermath of the event is that people want to help. Unfortunately, you have to be careful when it comes to volunteering, because people can get hurt if they try to do the cleanup themselves. It's unlikely they'll have the proper protective equipment or sufficient knowledge of what the potential health risks are. What's more, responders and scientists don't want the evidence to be taken—like a crime scene, a careful analysis of what's happening should be undertaken before the cleanup begins.

However, this doesn't change the fact that people in the affected community will want to help. The case in Sri Lanka was particularly frustrating to residents because of the strict COVID-19 lockdowns that were already in place at the time of the spill. Following the nurdle spill, Asha was able to shift the focus away from cleanup by establishing a crowd-sourced observation network. In a situation like this, where millions of nurdles were washing up on beaches up and down Sri Lanka's west coast, the most helpful thing for volunteers to do was to take pictures. We didn't just need pictures of the polluted beaches, we also needed pictures of the unpolluted beaches.

During an environmental crisis, the response team's most precious commodity is time. And they can lose a lot of it trying to figure out where they need to allocate their resources. Thus, documenting the locations where there is no observed pollution—on a real-time, daily basis—is hugely helpful, especially when there are limited resources. And these updates can be easily posted online through a platform like Twitter. Crowd-sourced photographs—which should be timestamped and georeferenced—are citizen-generated data points, and data points are the cornerstone of good science.

Empowering citizens to become scientists during an environmental crisis can be a game changer: not only are you helping people to contribute in a meaningful way, but you are also keeping them safe from harm. Additionally, in contributing to the process as a scientist, they become more invested in the scientific outcome.

Teamwork

The Sri Lankan fact sheet is important not just because of its impact. It is also important because it was the product of a team effort. While there were a number of scientists who contributed to the content, along with Ken Kostel, a WHOI staff science writer, the person who actually designed and published the final product was not a scientist at all, but a graphic artist by the name of Katherine Spencer-Joyce. If we hadn't worked with Katherine—if one of the researchers had come up with some impenetrable-looking graph instead, or tried to wrangle the page layout—the end result would have been unprofessional. Our fact sheet would have been just another boring, unread document in overflowing email inboxes everywhere.

In order to deliver content in a meaningful way, scientists ultimately have to partner with professionals in other fields. This bears repeating, because so often we try to save money by doing the work of others ourselves. But remember, you are *not* an artist. Graphic designers, in contrast, have the gifts and necessary training to translate complex ideas into visual images that are easy to grasp. As Donald Norman noted in his book *Emotional Design*:

Know your audience.

Figure 10.1 Take advantage of writers and artists to make your science more accessible. On the left is a fact sheet designed by an artist, on the right is one made by a scientist.

"Attractive things make people feel good, which in turn makes them think more creatively ... positive emotions are critical to learning, curiosity, and creative thought."[7]

Another obstacle that we need to overcome when working with interdisciplinary professionals is trust. We need to overcome our bias that you need a PhD and published papers to be an expert or to understand complex scientific ideas. We need to see communication as a team effort—and being a successful team player entails trusting the people you are working with, whether they are graphic artists, writers, your institution's PR team, or professional responders, like Steve Lehmann, who don't run their own labs.

In the end, communicating content is about more than the content. Above all, the finished product has to be eye catching. Looks do matter. While you may not be able to judge a book by its cover, you can definitely influence a book's sales, and target a particular market through that cover.

Every visual aspect of a well-designed fact sheet, magazine, book, website, or advertisement was the result of a conscious decision by a graphic designer. Margins, font type, graphics, color schemes, spacing, photographic touch-ups—all of these details add up to create overall visual appeal. Underlying messages can be skillfully and abstractly conveyed through patterns, illustrations, and cartoons.

In today's world of shrinking attention spans, good design is essential for communication. I always tell my students and colleagues: when you create some sort of figure, whether it be a graph or a photograph with a caption, your aim is for other people and publications to use the design you created.

And in the end, that's how I know our fact sheet was a success: because we heard from multiple stakeholders that it was useful.

The same goes for my personal lectures and presentations. Even when I am giving a simple PowerPoint presentation, I don't hesitate to have a graphic designer look over my slides. If I want to keep my audience engaged, if I don't want to miss the opportunity to communicate, then the slides need to be awesome. A great graphic is a powerful communication tool. Thus, the day you find a great graphic designer, make sure you continue to build and nourish that relationship, because they are invaluable communication partners.

And in building that relationship, you are on the road to establishing the key element of trust. At the beginning of any project, I usually just scribble out what I'm aiming for on a piece of paper, and then let whomever I'm working with do their thing. There's no micromanagement, second-guessing, or looking over a shoulder while they work. When it comes to design and visual communication, I am not the expert. I just wait to receive the final project, proofread it to make sure the information is accurate and there are no typos, and that's it. This level of trust may not come easily at first, which is why you want to build preexisting relationships. If you don't know where to start, then ask a colleague or friend for a recommendation. Make use of *their* preexisting relationships.

While I think there are plenty of scientists who recognize that professional graphic artists, web designers, writers, photographers, editors, and translators have skills that they do not have—even if the scientist in question has an artistic eye, a knack for writing, or can speak another language—when it comes to hiring these people, you have to make sure that you put your money where your mouth is and include them in the budget.

Even for the Sri Lankan fact sheets, which I knew would be important, I wound up having to ask my boss for extra cash, because we had no research money left over for the graphic artist. I'm lucky enough to have a great boss who was willing to pay extra. But there are plenty of other organizations that may not want to pay for freelancers. They might say: you can do this yourself, why do you need someone else? While that might be true—you *can* do it yourself—just keep that end result in mind: do you want all your hard work to go unread because it looks unprofessional, or do you want other organizations to use what your team made, because it is the reference *par excellence*?

Box 10.2 What the Experts Say: Graphic Artists Are Your Friends

In today's digital world, working with graphic artists has never been more important. Well-designed images not only increase the impact and accessibility of your findings, but also help improve your research.

The evidence isn't just anecdotal: a study published in 2019 paired six teams of researchers with graphic designers over a three-month period to explore the potential of interdisciplinary teamwork.[8] Each partnership was tasked with creating an infographic that would be presented at the 2019

American Association for the Advancement of Science (AAAS) annual meeting, covering such topics as plant-pollinator networks and global food systems.

The findings? Working with artists made the scientists better at what they do. Every step of the creative process, from identifying a target audience to deciding what message to convey in an infographic, resulted in the researchers pushing their "science forward." As the authors of the study wrote:

> In many cases the middle ground had to be found between the scientists' conviction that the graphics accurately and comprehensively represented the data, and the artists' emphasis on streamlining the messages to make them easier to understand. ... The process helped [the scientists] to identify the central components of their work and to note areas that they had not studied sufficiently.

The authors concluded their study by recommending that scientists not only work with graphic designers in order to make their work stand out, but also that they do so early on in the research process, as these collaborations will go on to "improve [the scientists'] current and potential future research."

Conclusion

You don't need to be an expert in string theory to know that it's impossible to predict how one event will influence another. When I answered Lihini's first email, in which she asked if I could help out with the *X-Press Pearl* chemical spill, I wasn't thinking about anything other than volunteering my time. I didn't foresee that I would wind up working with Asha. I didn't foresee that the US State Department would reach out to me—an independent scientist—because they were unsure of what, exactly, the proper steps in an environmental crisis response were. And that my answer would lead to me speaking to the US embassy in Sri Lanka on maritime disasters. And, most importantly, I didn't foresee that this spill would open up a new line of scientific inquiry on an important and understudied subject: nurdle spills.

At any point in the process, from answering Lihini's first email to publishing the fact sheet, I could have said, "I've done enough here," and gone back to my day-to-day work in the lab. But the satisfaction that comes from working with others and being open to communication in a crisis—not doomsday speculations in the media, but real data-driven contributions—can not only make a difference in the response itself, it can open your mind to new ideas. It can make you a better scientist.

In this case, we all realized that nurdles were scientifically interesting. Even though there are millions and millions of them floating around in marine ecosystems around the world, they are hard to research because no one knows when exactly a nurdle enters the water. When a nurdle washes up on a beach near you and you pick it up, can you tell how long it's been floating around out at sea? One year? Five years? Fifteen years? It's impossible to say.

This spill was different, though. We knew precisely when and where the nurdle spill took place. We knew exactly what had happened to the nurdles—they had been burned in a fire. Asha was able to collect samples as soon as they started washing up on the beaches. We published a data-driven fact sheet on the nurdles in just ten days—that was the scientific response part of the equation. We then published a joint peer-reviewed paper on the nurdles five months later;[9] extremely fast by scientific standards, but considerably slower than the time needed to publish and distribute a fact sheet. The different cultures of responders and academia translated into different needs, time scales, and value systems. Both have their place.

And now, a recent spill of unburnt nurdles near the port of New Orleans has caught the attention of Bryan James (a postdoc in my lab). Before, this event might have slipped by without anyone giving it a second thought. But after having just completed a study of burnt nurdles, we knew that it would be equally useful to complete a study of unburnt, pristine nurdles. And because our scientific antennae were already up, we were able to obtain a sample from this spill.

What was originally deemed by some to be obvious and uninteresting has gone on to form the basis of a new line of scientific inquiry. And it all happened because we worked together. We were a team.

Notes

1 Tim Lydon, "3,000 Shipping Containers Fell Into the Pacific Ocean Last Winter," *Revelator*, June 2, 2021, https://therevelator.org/container-ship-accidents.

2 Asha De Vos et al., "The M/V X-Press Pearl Nurdle Spill: Contamination of Burnt Plastic and Unburnt Nurdles along Sri Lanka's Beaches," *American Chemical Society* (November 29, 2021): https://doi.org/10.1021/acsenvironau.1c00031; X-Press Pearl Incident Information Centre, www.x-presspearl-informationcentre.com.

3 www.nurdlehunt.org.uk.

4 www.whoi.edu/news-insights/content/sri-lanka-faq. For more on the importance of assigning proper credit when dealing with international crises and research topics, see de Vos, "The Problem of 'Colonial Science,'" *Scientific American*, July 1, 2020, www.scientificamerican.com/article/the-problem-of-colonial-science.

5 www.whoi.edu/wp-content/uploads/2021/06/Fact-Sheet-XPressPearlSpill24Jun.pdf.

6 To her credit, Asha sent the fact sheet to her contacts in the Sri Lankan government before it went live so that there were no surprises with regard to the information that was going to be made public.

7 Norman, *Emotional Design* (New York: Basic Books, 2003), chapter 1.

8 C. K. Khoury et al., "Science–Graphic Art Partnerships to Increase Research Impact," *Communications Biology* 2, no. 295 (2019): https://doi.org/10.1038/s42003-019-0516-1.

9 Asha De Vos et al., "The M/V X-Press Pearl Nurdle Spill: Contamination of Burnt Plastic and Unburnt Nurdles along Sri Lanka's Beaches," *American Chemical Society* (November 29, 2021): https://doi.org/10.1021/acsenvironau.1c00031.

11 How Communication Can Make You a Better Scientist

Thus far, we've mostly discussed the challenges that scientists face when trying to communicate with different stakeholder groups. However, there is another facet of communication, which we don't hear often enough: it can make you a better scientist.

In practicing your communication skills, you will gain a greater understanding of what other stakeholder groups value in the scientific community. Nonscientists may pose thought-provoking questions that go on to influence your research. And you may learn some valuable lessons while working with others that you then incorporate into your own culture.

This is a good time to remind ourselves that communication takes practice. Just like science, it is incremental. It is a set of skills that you have to keep working on, in the same way a runner trains for a marathon: some days you focus on strength training, some days you do short runs, while other days you do long runs. The point is, you don't train for a marathon by running a marathon every day.

Not only do you need practice, you also need to deepen your understanding of the way the world works. How is government policy formed? How does a community react to a crisis? How does the media cover a crisis?

These questions should be the first step in learning how to communicate science outside the ivory tower. And as with so many things, the answers to these questions can be found just outside your doorstep, in your local community. Coincidentally, your local community is also the best place for you to work on developing—and maintaining—your communication skills.

So, just how can local engagement make you a better scientist? Let's take a look.

Putting Your Finger on the Pulse: Town Council and State Legislature Meetings

In addition to reading your local paper, town council and school committee meetings are a good place to get a feel for what's happening in your community. They can also be an ideal forum for you to talk about an issue in which you have some sort of special knowledge or expertise.

DOI: 10.4324/9781003341871-15

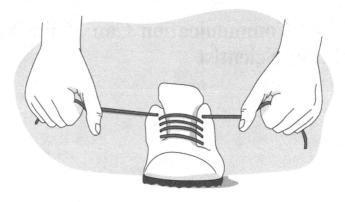

Learn to walk before you can run.

Figure 11.1 You have to train, practice, and strive to continually improve your science communication skills. Three hours of training will not make you a world-class communicator, but it will put you on the right track.

One topic that has long interested me is plastic. As we all know, plastic is an integral part of the modern world. Packaging, toys, shopping bags, computer parts, food containers, plumbing, appliances ... so many pieces of our lives—big and small—depend on plastic. Eventually, most of these items are discarded, recycled (only 9% of all plastics in 2015[1]), or lost. Sometimes they wind up in dumps. But sometimes they become pollution. Pollution that makes its way into a nearby stream, which makes its way into a river, which makes its way into the ocean. Or sometimes plastic enters the ocean directly, whether in a shipping accident, like with the *X-Press Pearl*, or through illegal dumping.

As marine pollution, plastic comes in a variety of forms: nurdles, bags, wrappers, bottles, caps, and so on. Up until recently, it was assumed that plastic was inert—in 2016, for example, the UN Environmental Programme stated that Styrofoam would last in the environment for thousands of years.[2] The pervading belief was that the chemical composition of plastic didn't break down over time. Instead, it just became smaller and smaller, gradually becoming what's known as microplastic.

Plastic pollution, like ocean acidification, is what might be called an existential crisis. It is not a one-time event, but an ongoing threat that will likely become increasingly worse with time. As stated in *Science*, plastic waste is estimated to rise from six billion metric tons today to greater than twenty-five billion metric tons by 2050.[3]

Many citizens are concerned. One of the more popular policies to curb such pollution has been for local communities to ban plastic bags. But how much do we really know about plastic bags? Does enacting such bans have an effect on the total amount of plastic waste entering the oceans? Do plastic bags really last forever, as so many people assume? Given the ubiquity and convenience of plastic in the modern world, wouldn't a better solution be to design next-generation plastics that degrade once they have served their purpose? Or is enacting a plastic bag ban just an easy way to make people feel better without actually addressing more pressing environmental issues?

I began researching plastics in 2008. In fact, it was only because of *Deepwater Horizon* that I went back to studying oil spills. So I already knew that, when it comes to plastic pollution, there isn't a lot of data on the short- and long-term fate of plastics once in the ocean.

In 2013, plastics reappeared on my scientific radar when I heard that the Rhode Island legislature was considering a statewide ban of plastic bags. Curious, I attended a public hearing in order to provide a testimony. I thought the day was going to be similar to a federal hearing in Congress—green felt table, carefully vetted speakers who present contrasting arguments, lots of pomp and circumstance. But when I showed up at the state legislature, I found that I had to sign my name to a list that already had two dozen signatures on it.

During the hearing, speaker after speaker stood up and argued passionately against plastic bags. Of course, they all had a point: single-use plastic bags are hardly environmentally friendly—no sooner have you brought one home then it becomes waste. But, here's the thing. None of these speakers had any data to back up their claims.

When it was my turn to speak, I reminded the legislators of this fact. If Rhode Island really wants to do this right, I said, we should first gather evidence, and then form a policy based on what the data tells us. There are just too many variables when it comes to plastic pollution to know what sort of impact a bag ban will have. In terms of local ocean pollution, it's likely that nitrogen runoff from golf courses and lawns is a considerably greater threat to Rhode Island's Narragansett Bay than plastic bags. Local scientists such as Robinson Fulweiler had already researched this particular subject, and it was a much more tangible problem with measurable impacts on the local economy.

We have to be careful, I went on. There is a finite interest in changing policy with regard to the environment. If you can only win a select number of battles, wouldn't it be better to address the issues where we have data to back up our claims, and where we know our impact will be greatest?

To be clear, I wasn't arguing in favor of plastic bags. I was just saying, if we're going to do this, let's do it right. Let's do it in a way so that we're sure we're making a difference.

Of course, this sort of approach—talking about data and analysis—works when you are talking with other scientists. But it doesn't always reach the concerned public. There's a popular saying when it comes to policy debate: if you want to reach people, focus first on their belly, then on their wallet, and last on their brain—because intellectual arguments are the least effective.

Scientists and environmental activists are often on the same side when it comes to policy. But sometimes, the concerned public becomes so emotionally invested in a cause that they don't take into account all of the nuances. In this case, every single one of the people who testified ahead me thought I had been hired as a pro-industry shill—they thought I was taking an anti-environmental stance. Nothing could be further from the truth. I was just saying we can't create policy based on the eye test. What's the point of doing something that makes us feel good if we can't measure how much good it is truly doing?

While this particular example did not lead to a successful outcome in the short term—in fact, for years afterward I was misquoted during plastic bag debates in other communities—it did affect my research. It made me a better scientist. The proposed ban raised important questions, and this led me—and others—to try to answer these questions in the lab. When you engage with an issue at the community level, you might find you don't have an immediate answer. But that's okay. Science takes a long time.

It was clear that plastic pollution was both an issue that people felt strongly about and one that needed to be addressed scientifically. And one of the new research questions it raised was: how long does plastic really last in the environment, anyway?

Box 11.1 What the Experts Say: Social Media Success

Anyone who has ever had a brush with fame knows that celebrity status changes you. No, you didn't accidentally stumble onto a TMZ.com story: this phenomenon affects scientists too. Researchers Donna Rockwell and David Giles broke fame down into a four-phase process: (1) love/hate feelings; (2) addiction; (3) acceptance; and (4) adaptation. In other words, becoming famous affects your behavior. One of the most striking changes that takes place is the following: if "happiness becomes synonymous with celebrity ... one incessantly strives to remain the center of adoring attention."[4] None other than Einstein once declared: "With fame I become more and more stupid."[5]

In contrast, however, a recent study of 17,157 scientists on Twitter documented their online behavior over the course of three years.[6] Of these, about 4% (670) experienced an unexpected "microviral" event at some point, which consequently changed the way they tweeted. These scientists began posting more often, narrowed their topics of interests, stuck more closely to posts that resembled their viral tweet, and gained more followers. Interestingly, the study also found that these scientists' tweets started to become more positive, which points to a change where a scientist recognizes that positive, objective tweets are more likely to gain followers and generate a greater impact.

Expressing an Opinion

In terms of low-risk engagement in a community matter, posting on social media or writing a letter to the editor (LTE) of your local paper can be a good first step. Both these formats help you practice big-picture thinking. How do you organize your thoughts so that you can convey your message— your *so what*—in only 280 characters, or 150 to two hundred words?

Of course, while the short-length format is similar, there is an important difference: posting on social media is much easier than publishing an LTE. For the former, all you have to do is click. A newspaper, meanwhile, has to deem your LTE engaging and pertinent enough to be accepted.

Thus, if social media is a good place to hone your writing chops, an LTE is a good place to practice the submissions process. Writing an LTE is a good exercise, and I encourage you to try it sooner rather than later. It can be as simple as scribbling out a draft while reading the news over breakfast.

In order for a letter to the editor to be published, you should be sure you're doing one of three things: taking a clear stance on a community issue, stating an opposing viewpoint to a recently published article, or pointing out a mistake in a recently published article.

Newspapers get lots of LTEs, so there's a good chance you may be rejected. But while a rejection (or, even worse, no response at all) may make you feel like you've wasted your time, in fact, this writing exercise is just that: an exercise. Writing a letter to the editor is a way for you to work out your communication muscles. To make them stronger. As long as you keep it up, that training will eventually reap rewards.[7]

Box 11.2 Rejected

On this note, it's important to mention rejection. Rejection is a part of the publishing process. Stephen King's first novel *Carrie*? Rejected by thirty publishers. *Chicken Soup for the Soul*? Rejected 144 times! Yes, just like James Joyce, Madeleine L'Engle, Dr. Seuss, Agatha Christie, John Grisham, Walt Whitman (self-published), Marcel Proust (also self-published), and many, many others, you too will be rejected. The difference between scientific publishing and mass-market publishing is that the editors of newspapers, magazines, and books will rarely explain themselves—if they even deign to acknowledge your existence. Scientists expect feedback, so the silent treatment can come as a real culture shock. We all know that if a scientific journal rejects a paper following review, you will get detailed feedback of some kind. You might not agree with the outcome, but you can see what your peer reviewers took issue with, and what you need to improve. You can (somewhat) rationalize the rejection.

Newspaper and book editors, on the other hand, may not tell you anything at all. Just, "Sorry, we're not interested." Or, "Sounds interesting. Let me look at it and get back to you in a few days." A few days come and go.

No word from the editor. Did they forget about you? Or were you put on the back burner because a more pressing story just broke? There's no way around it: rejection will always hurt.

But this is how the industry works. No editor is obligated to tell you why they rejected your article. No editor is obligated to follow up at all. The single most important thing I can say about rejection is: *do not take it personally*. Do not write a rebuttal of any kind, or dwell upon feelings of ill-will toward a specific individual or publication. Ultimately, you are trying to cultivate a long-term relationship; if you act unprofessionally, that door will be closed to you forever. Editors are overworked and underpaid. They have to deal with hundreds of pitches just like yours. There is no way for them to respond to each one.

However, we all know that feedback is invaluable when it comes to improvement. Just because you can't get feedback from the person who rejected your work doesn't mean that you can't get feedback at all. If you have a communications department, ask for their advice. Talk it over with your lab group or department. See if they have suggestions. Alternatively, hire a professional writer or editor and ask for a critique. As long as your article or proposal isn't *War and Peace*, a professional critique can be a relatively inexpensive way to gain insight into your writing.

Similar to an LTE, but much harder to write, is an op-ed. Op-eds are longer—usually from five hundred to seven hundred words—which leaves you enough room to back up your opinion. If you think in terms of paragraphs, a good rule of thumb is to aim for an introduction, three supporting arguments, and a conclusion.

Remember, brevity is your friend. Focus on one main topic. Catch your reader's attention with a good hook. You are not writing for other academics, so leave out overwritten, run-on sentences. And, even if you've heard it a million times, it bears repeating: don't use jargon. Finally, a scientist's op-ed usually addresses the intersection of public policy and science, so make it clear what your opinion on the policy is. Do you support it or not? What's your solution? It's not easy to publish an op-ed, but like a letter to the editor, writing one is good practice.

After my testimony in the Rhode Island legislature in 2013, the issue of banning plastic bags came up again in Boston in 2017. This time, rather than get involved at the government level, I decided to submit an op-ed to the *Huffington Post*. Unlike my testimony at the state legislature, I was not speaking to policymakers and environmental activists. I was speaking to the general public.

I had more time to reflect on why my first engagement with a plastic bag ban hadn't gone over well. But, while I changed my tone somewhat, I didn't back down from my core message, which was: we still didn't have the data we needed on plastic pollution. In this opinion piece, I wrote:

We lack a comprehensive and rigorous scientific understanding on the relative sources of plastics that reach the coastal and open ocean; how they behave, react, and degrade; how far they travel from land and, if and when, they sink; and the injuries they impact at each and every level of the entire ecosystem. To gain that understanding requires an investment in additional research. If society really wants to implement effective public policies that make progress toward reducing this major problem at all levels—locally, nationally, and globally—it will have to devote more time and resources to find them.

As plastic bag banning proposals continue to pop up in coastal communities across the country, I wonder if we will finally decide to devote the resources needed to meaningfully answer the question about plastic bags' impact on the environment. And while plastic pollution is also a visible problem, will we devote as much passion and energy to less visible marine pollution issues where the science, in fact, is clear?[8]

Why isn't there more data on plastic pollution? Clearly, many members of the public are extremely concerned about the issue. Part of the reason is because plastic is visible. If carbon dioxide were black and nitrogen fluorescent green, you can bet that a lot more people would be concerned about global warming and nitrogen runoff. It's that simple.

However, that argument doesn't answer why scientific research into plastics is lagging. My own interest in this question—how plastics behave, react, and degrade in the ocean—and the obvious lack of scientific data to support public policy, eventually led to more research on the subject.

In 2019, my colleague Collin Ward and I published a paper with several other researchers on polystyrene.[9] Polystyrene accounts for 6% of the current global plastic market, and is found in food containers, packaging, and building materials. This was one of the first studies to show that these compounds break down much more quickly than previously thought, and that sunlight exposure is the key factor in their transformation to carbon dioxide and dissolved organic carbon. Essentially, we demonstrated that, contrary to popular belief, plastics don't last forever.[10]

Following this study, Collin decided to take a deeper dive into the matter. He was sure that at least some other scientists had done research on plastic bags.[11] Like all of us, he had seen the many posters and infographics that tell us how long plastic bags, bottle caps, straws, and so forth persist in the environment. This sort of information can be found in a variety of places: textbooks, lesson plans, government publications, nonprofit flyers and websites, in the media, and on public posters in parks and beaches. So Collin began a fact-finding mission: where were all of these estimates coming from?

Together with a summer student, the two of them spent months investigating the subject. And what did they find out? After researching fifty-seven plastics posters and documents—collected from thirteen countries across four continents—*there wasn't any peer-reviewed supporting data. At all.* Twenty-one of the fifty-seven graphics had no source. Of the others, all citations could be traced back to three original sources, none of which was based on peer-reviewed science. During a phone conversation with one of these sources, the person confirmed that the estimates came from volunteers, and there was no science behind them.[12]

And thus, we went on to do more research.

So What?

The next paper to follow in this line of inquiry was the one that brought it full circle. In September 2021, graduate student Anna Walsh, whom Collin and I coadvised, published a paper on plastic bags. In this follow-up study, she took single-use bags from three sources—CVS, Target, and Wal-Mart—and examined how each degraded following exposure to sunlight. The takeaway from her study was that even though each bag looked the same, they all behaved differently based on their additives. The bag from Target broke down into about 5,000 dissolved organic carbon compounds, the bag from CVS broke down into 13,000 formulas, while the bag from Wal-Mart broke down into 15,000 formulas.[13]

One constant that was revealed was that titanium dioxide, which is added to make things white, acts as a catcher's mitt for sunlight, and this helped to break down the plastic faster. So how long do plastic bags last before they degrade? The answer is: it depends. But, it's fair to say that when exposed to sunlight, they do not take five hundred years to decompose, as the local government of Nantucket reported.[14]

So what does this mean for the future of plastic? That the lifetime of a plastic product—its degradability—depends on its composition. We can, and should, try to design more environmentally friendly bags based on the information we learned in studies such as this one.

Just after this paper was published, we received a call from a *Politico* reporter, who saw our press release and wanted to interview Anna. She was interested and wanted to engage; however, as a grad student with limited media training, she was understandably reluctant to sit down with a journalist. But unlike my graduate advisor, who once told me, "Whatever you do, don't talk to the media," Collin and I recognized that there was a tremendous learning opportunity here.

And so, after Collin and I had a chat, I offered to Anna:

> I'll give the interview, but you should sit in and listen. If you want to jump in, go for it. After the interview is over, I want you to summarize your notes. What kinds of questions were you expecting the reporter to ask? How did we answer these questions? What was the most interesting moment? Did anything surprise you? And so on.

After the piece was published, Collin and I gave her a second assignment: write a summary of how the article played out relative to her expectations as the interview was taking place. We then had her send both synopses to the journalist.[15]

I now do this every time a suitable interview comes along, because there is simply no substitute for a student to see how the sausage is made. It's especially useful when the student is implicated in the research. They already understand the science, and now they get to see how those results are translated to a general audience via the media. They learn how to research a reporter ahead of time so that they'll have a better idea of whom they'll be dealing with. They learn how to take the time to prepare a message beforehand. Amazingly enough, I know plenty of scientists who will spend hours and hours getting a ten-minute presentation ready for a dozen colleagues. But when it comes time to talk to the media? Five minutes of prep time max. It doesn't work.

Afterward, when Anna told me how stressed out she would have been if she had had to give the interview, I told her:

> It's okay. Interviews are manageable because you should always have a list of questions from the reporter ahead of time. You should already know what you're going to say before the interview takes place. Even if you don't get a list of questions, you still know what a reporter is going to ask you: *so what?* She isn't going to ask about our radiocarbon analysis or the statistical methods we used. You and I might find that interesting, but she doesn't care about that. Her job is to tell the public why this matters.

For me, this type of training is so much better than mock interviews or potentially embarrassing moments where you have to stand up in front of a room of strangers and talk about your research. While useful, this type of training is simply less realistic, because your name and research aren't actually on the line. There's no substitute for the real thing.

So what? In the end, this is what it all comes down to. In 2013, I gave a testimony at a Rhode Island state legislature hearing on plastic bags. I attended this hearing because I was curious about the local environmental policy in my home state. At the time, I stated that we simply didn't have the data to know if a bag ban was an effective solution to stemming the rising tide of plastic pollution. As the years passed, this testimony, and the questions that it raised, grew into a veritable line of scientific inquiry that my colleagues and I pursued in the lab. It took eight years for us to finally publish a paper that dealt specifically with the subject that originally sparked my interest. This is a reminder: science takes time. Even in a crisis.

Did we conclusively answer the question as to whether plastic bag bans are effective? No, but we added our own puzzle pieces. That initial engagement in my local community made me a better scientist. I was able to align my research interests with a topic that I knew mattered to the people around me.

In the three years since our paper on sunlight and polystyrene was published, it's been cited over one hundred times. That tells me something. There are other scientists out there who are working on a similar line of research. They too will add their own puzzle pieces. Little by little, we will slowly complete the picture. Eventually, we will have better, more environmentally friendly material. No single person will be responsible for the breakthrough, because that's not how science works. Each scientist is building on the work of someone else. But one day, we will get there, the only way we know how: together.

Notes

1 R. Geyer, J. R. Jambeck, K. L. Law, "Production, Use, and Fate of All Plastics Ever Made," *Science Advances* 3, no. 7 (2017): https://doi.org/10.1126/sciadv.1700782.

2 UNEP, "Marine Plastics Debris and Microplastics: Global Lessons and Research to Inspire Action and Guide Policy Change," *United Nations Environment Programme*, 2016, https://wedocs.unep.org/handle/20.500.11822/7720.

3 R. Geyer, J. Jambeck, and K. L. Law, "Production, Use, and Fate." See also C. Ward and C. Reddy, "We Need Better Data about the Environmental Persistence of Plastic Goods," *PNAS* 117, no. 26 (2020): https://doi.org/10.1073/pnas.2008009117.

4 D. Rockwell and D. Giles, "Being a Celebrity: A Phenomenology of Fame," *Journal of Phenomenological Psychology* 40, no. 2 (2009): https://doi.org/10.1163/004726609X12482630041889.

5 P. D. Smith, "With Fame I Become More Stupid," *Guardian*, August 31, 2002, www.theguardian.com/books/2002/aug/31/biography.highereducation.

6 R. Hasan et al., "The Impact of Viral Posts on Visibility and Behavior of Professionals: A Longitudinal Study of Scientists on Twitter," *Proceedings of the Sixteenth International AAAI Conference on Web and Social Media* 16, no. 1 (2022): 323–34.

7 It is worth noting that outside of the big national papers, a well-written science opinion piece has good odds of being accepted for publication.

8 C. Reddy, "To Ban or Not to Ban Plastic Bags? That's Not the Best Question," *Huffington Post*, March 28, 2017, www.huffpost.com/entry/to-ban-or-not-to-ban-plastic-bags-thats-not-the-best_b_58da724ae4b0e6062d923105.

9 C. Ward, C. Armstrong, A. Walsh, J. Jackson, and C. Reddy, "Sunlight Converts Polystyrene to Carbon Dioxide and Dissolved Organic Carbon," *Environmental Science & Technology Letters* 6, no. 11 (2019): https://doi.org/10.1021/acs.estlett.9b00532.

10 We showed that the polystyrene does not disappear by breaking into small pieces (as if it had been cut with mechanical scissors). Instead, sunlight leads to chemical degradation—the equivalent of chemical scissors breaking the very structure of the plastic.

11 Collin was not alone in assuming that there was valid scientific data behind these infographics. After the results of his study were published, a former postdoc of mine, "New" Bob Swarthout, called me to say he had once assigned this very fact-finding mission to one of his undergrads. When his undergrad returned and said there wasn't any data, he thought that his student was a slacker and didn't want to do the work.

12 C. Ward and C. Reddy, "We Need Better Data about the Environmental Persistence of Plastic Goods," *PNAS* 117, no. 26 (2020), https://doi.org/10.1073/pnas.2008009117. Collin, who is untenured, showed a lot of courage in publishing this piece. He also did something very smart: he reached out to the groups who could have caught blowback and let them know what was coming. This is a good time to remember: nobody likes surprises.

13 A. Walsh, C. Reddy, S. Niles, A. McKenna, C. Hansel, and C. Ward, "Plastic Formulation Is an Emerging Control of Its Photochemical Fate in the Ocean," *Environmental Science and Technology* 55, no. 18 (2021), https://doi.org/10.1021/acs.est.1c02272.

14 Town and County of Nantucket, "Stop the Straw," 2018, www.nantucket-ma.gov/1131/Stop-The-Straw.

15 Note that we also reached out to the reporter before the interview to make sure she was on board with the process. In an email, I wrote: "Would you mind playing a role in training our next generation of scientists? ... Seems like a win-win. Anna gets experience; Collin and I get a chance to provide an exercise that we cannot perform in our labs; and you and your colleagues in the media will have scientists better prepared for media interviews in the future."

Conclusion

The Pathway to Success

Adam McKay's film *Don't Look Up* struck a chord with many climate scientists. The satire, which tells the story of two astronomers whose discovery of an impending disaster is met with a worldwide shrug ("You discovered a comet? Good for you!"), captured the frustration many of us feel when trying to communicate important scientific findings.

But even if it sometimes feels like the world just doesn't care, it is worth remembering: scientists have previously identified and changed the course of existential threats before. We have success stories to build on. And we should go back to these stories from time to time, in order to remind ourselves that we have the capacity, the courage, and the patience to enact real change.

One such story is that of Clair Patterson (1922–95), the unheralded geochemist from Iowa who placed the missing puzzle piece that helped scientists finally answer the question: how old is the Earth? Amazingly enough, Patterson's estimate of 4.55 billion years, first published in 1956, still stands today.[1]

In order to arrive at his measurement, Patterson worked with the ratio of uranium to lead (U–Pb dating)—first in zircon crystals, then in ancient iron meteorites—using the half-life of uranium to accurately wind back the clock billions of years. But as he was undertaking his early lab work at the University of Chicago, Patterson noticed something. All of his zircon crystals were becoming contaminated with unusually high levels of atmospheric lead.

At first, this was no more than an obstacle to overcome. The immediate outcome of the contamination was that Patterson had to develop many of the super-clean lab techniques that researchers today take for granted. But it also posed a question that stuck in his mind: where was all this lead coming from?

Getting the Lead Out

In the mid-twentieth century, lead was everywhere. It was used to seal canned food. It was in paint. The pipes that carried drinking water to our homes were—and in some cases, still are—made out of it. Dinner plates were coated in lead glaze. Even tubes of toothpaste contained lead. Considering how toxic it is, it's amazing how much lead there was in our day-to-day lives. But when it came to atmospheric lead pollution, all of these examples were no more

DOI: 10.4324/9781003341871-16

than a drop in the bucket. There was one major driver of global lead poisoning—the kind that can be inhaled directly into our lungs—and it was this: leaded gasoline.

In 1921, the inventor and engineer Thomas Midgley discovered that adding tetraethyl lead to gasoline would reduce engine knock. The best thing since sliced bread? Not quite. The result of this additive was that organic lead was now being pumped straight into the atmosphere, and worse yet, it was now in a form that was easily absorbed by mammals everywhere. The toxicity of the tetraethyl lead additive was known from the very beginning; the first manufacturing plants run by the Ethyl Group made the news in 1924 following mass lead poisoning incidents and deaths among its workers.[2] Midgley himself became quite sick after working with the compound. But leaded gasoline was simply too profitable, and too convenient, to replace.

And to be sure, few people even conceived of the possibility that car exhaust would go on to change the composition of the Earth's atmosphere—until Patterson. What would have happened if he hadn't noticed that lead particles in the air were contaminating his lab? How high would the atmospheric concentrations of lead have had to have reached before anyone noticed that we were poisoning the entire planet with a neurotoxin?

After Patterson's big announcement about the age of the Earth, he turned his attention to the problem that had been nagging him ever since he first began his research using U–Pb dating: why was the atmosphere full of lead particles? Was it a natural phenomenon? Or anthropogenic? Of course, he had a hunch, but it wasn't until he traveled to Greenland to extract ice cores that he was able to prove his hypothesis. These ice cores revealed that small quantities of lead first began to appear in the atmosphere during the Industrial Revolution. And then, in the 1920s, the parts per million skyrocketed, and were continuing to increase with each passing year.

Sound familiar? Yes, Patterson was the first to recognize that Greenland's ice sheets held the clues to analyzing the historic composition of the atmosphere. And just like climate researchers today, Patterson's discovery was not met with open arms. Following the publication of *Contaminated and Natural Lead Environments of Man* (1964), Patterson immediately lost funding for his work. The Ethyl Corporation tried to pressure Cal Tech into firing him. And he even lost the support of some of his colleagues, who believed that matters of public health weren't "real" science. After all, Patterson was part of the elite group of postwar scientists at the University of Chicago. His contemporaries won Nobel Prizes in physics and chemistry. Patterson, meanwhile, had gone from the grandness and mystery of the formation of the solar system to spending his life researching gasoline pollution. In terms of prestige … it wasn't the same.

Furthermore, he insisted that virtually every other scientist working with lead was using faulty data and publishing inaccurate results—he knew that if they weren't using super-clean lab techniques, there was no way to avoid contamination. Such statements (which, of course, were true) did not make him popular among his colleagues.

But despite the challenges he faced—the risks to his career, his reputation, his research funding, and his legacy—Patterson continued his campaign to educate the public, policymakers, and other scientists, notably in the field of public health. He knew it was imperative that we remove lead from gasoline. And he succeeded. In 1970, the United States passed the Clean Air Act. In 1973, Congress banned toys with high levels of lead. In 1974, it passed the Safe Drinking Water Act. In 1975, unleaded gasoline was introduced.[3] In 1977, the government banned the use of lead in indoor house paint. And in 1986 lead was phased out of all plumbing materials, from pipes to solder.[4] The impacts were real. According to a recent study, between 1976 and 2016 the mean blood level of lead in the United States dropped from "12.8 to 0.82 µg/dL, a decline of 93.6%. ... There is no safe level of lead exposure, and child BLLs [blood lead levels] *less than 10 µg/dL* are known to adversely affect IQ and behavior."[5]

Again, we have to ask ourselves: what would have happened if Patterson had been unsuccessful in his campaign? If he were alive today, would his message have gotten through to policymakers? In 2014, I received the Clair C. Patterson Award for my research on oil spills. In my acceptance speech, I said:

> Patterson is an example to all scientists that we should not cloister our-selves in our science, and that sometimes we should be willing to follow our scientific noses, step beyond our comfort zones, and explore new areas, even if that comes with the price of criticism and persecution. Patterson's life is the ultimate parable illustrating how basic science rewards society. If he had not pursued the basic question about the age of the Earth, he would not have been able to change worldwide policies that diminished harmful exposure of lead.

Despite all his contributions—and determining the age of our solar system is a truly remarkable feat—Clair Patterson remains relatively unknown today, even among academics. When I first read about him, he made me appreciate how lucky I am to be a scientist. He also made me appreciate the impact of suc-cessful communication. To wit: even though nearly half of all Americans would dispute his calculation for the age of the Earth (four out of ten adults surveyed in the United States thought the Earth was closer to 10,000 years old[6]), few would deny what is known about the negative impacts of lead pollution.

Graduate students, postdocs, and assistant professors are generally the most excited about venturing outside of our peer group. You are young. You are concerned and want to make a difference. You have the energy needed to change the status quo. But where do you begin?

The most important piece of advice I can give is to be patient and start small. You need to accumulate knowledge and build your reputations as sci-entists. You need to figure out how the world works. And above all, you need to practice your communication skills. You can't take a single three-hour training program, conclude with a mock interview, and then say, "Okay, let's do this." Or, to put it another way, passing your driver's license test does not make you an experienced driver. Like it or not, the *New York Times* is unlikely to seek out graduate students for interviews. A US Senate committee will not ask a junior scientist to testify.[7]

But this doesn't mean that you can't gain experience. You just have to be strategic in how you go about it. Let's recap the ten challenges that scientists face when communicating.

1 Career: Is the risk worth the reward?
2 Culture: Every stakeholder group is different.
3 Network: You can't exchange business cards during a crisis.
4 Speaking Out: Public statements have real-world consequences.
5 Competition: It's easy to lose sight of the big picture.
6 Process: Science is a never-ending jigsaw puzzle.
7 Misinformation: How do we counter scientific falsehoods?
8 Legal: How do we deal with subpoenas and other forms of harassment and intimidation?
9 Impact: How can you make every encounter a positive one?
10 Teamwork: Communication is a team sport.

To these challenges, I'd now like to add a list of solutions. Of course, there is no single way to overcome each of these challenges, because for each person, for each situation, they will be different. Instead, think of this list as a starting point.

1. Career: Start Local, and Be Patient

Everyone starts at the ground level. And this is a good thing, because it allows you to practice communication with minimal risk. Remember, practicing communication is like training for a marathon—you have to keep doing it, week in and week out. Part of successful communication is figuring out what sort of situations you enjoy and what comes naturally to you. Learn to play to your strengths, recognize your weaknesses, and focus on what to improve. Consider the following starting points:

- Follow local news and connect each topic to your specialty. Educate yourself: read newspapers, listen to radio programs. Consider other points of view. Hone your skills by critiquing whether these science pieces are clear and accurate.
- Attend town or city council meetings. Is there an issue in which you can weigh in as an expert?
- Visit the office of your local, state, and federal legislators. You may not meet the elected officials but, as a constituent, you will be heard.
- Meet people outside your lab group. Talk to both administrative professionals and other academics (political science, communications, graphic

design, marketing, drama, journalism, creative writing) at your organization. Find out what people think science can and can't give. Be wary of "science" stereotypes and strive to overcome them.

- Get involved with local organizations, such as church, library, and school groups. Participate in science and Maker fairs. What do these groups want from a scientist? This is an attainable, low-risk way to interact with the concerned public and future scientists and engineers.

2. Culture: Know Your Audience's Comfort Food—and Make Sure You Serve It to Them

Too often, we focus only on the language that we use, the words that we say. But successful communication depends on so much more than our words. It's also how—and when, and to whom—we say it.

Let's say you aced your AP French exam in high school, but you've never actually lived in France. You make your first trip to Paris. How quickly would it take a French person to identify you as an outsider? Even if you can read *Le Monde* and hold your own in a conversation, you would probably be identified as a foreigner the second you stepped off the plane. All those other nuances—body language, dress, personal space, and so on—would give you away immediately. Communicating with another stakeholder group is so much more than just knowing the language. It is understanding the aspects of culture that are unspoken: the values, the habits, the expectations, the definitions of success.

Have respect when venturing outside your peer group. Remember that you are the outsider. You are the student. Just like the late chef, author, and TV host Anthony Bourdain in *Parts Unknown*, you have to listen to—and learn from—what the locals are telling you.

3. Network: You Don't Have to Like the People You Work With, but You Do Have to Earn Their Trust

It took years for Steve Lehmann and me to become friends. But whether or not we were friends wasn't the issue. What was important in our relationship was that we learned to have mutual respect. That we trusted each other. And the only way to acquire trust is to earn it. That's why you have to start building your network in tandem with your career. You have to prove to people outside your field that you know how to be a team player, that you are competent and reliable. That you are able to value the experience that they bring to the table, even though they might not have a PhD. Trust is the secret ingredient that allows you to have a dissenting voice that is taken seriously, rather than being perceived as a mere annoyance.

Start building your network by reaching out to local politicians, non-governmental organizations, journalists, responders, government labs, and environmental organizations. What do you have to offer? And what can they teach you? You are striving to create long-term relationships with every person you meet. There are no one-offs.

4. Speaking Out: It's Not What You Say, but Understanding What People Are Going to Do with the Information after They Get It

What is the role of a scientist in a crisis? Whistleblower? Soothsayer? Datacruncher? Authority? Sage? Part of the reason we have to choose our words carefully—along with our target audience—is because of unintended consequences. When a reporter asked me if the oil spilled in Buzzard's Bay in 2003 contained naphthalene, I confirmed that it did. But I didn't anticipate where he was going with that line of questioning. Naphthalene was not the most pressing issue in that spill, so I didn't think to clarify, "Naphthalene is relatively short lived in the environment and is only toxic in water in high concentrations." And even if I did add that clarification, would he have included it in his article? His story angle was already determined before the interview.

A much better answer would have drawn the focus away from the scientific details. Members of the affected public do not care about naphthalene per se. What they want to know is: how is this going to affect *me*? When can I take my fishing boat back out into the bay? What is going to happen to the wildlife? What is going to happen to the beaches? When will seafood be safe to eat?

Prior to the interview, I should have recognized this and come up with a simple message that addressed these concerns:

> Yes, there are some unfriendly compounds out there, but they are not a big deal. There is going to be some damage to birds, some damage to salt marshes, but I imagine that once the data comes back, all the fisheries are going to be open. Based on my past experiences, this is not going to be the end of the world.

Of course, anticipating unintended consequences is a tall order. You'll never be able to foresee everything that can go wrong, or the various ways in which your words might be misinterpreted. But being mindful of the major issues at play, and crafting your message appropriately, will certainly help mitigate any unforced errors on the part of science.

5. Competition: Don't Get So Wrapped Up in Your Career that You Lose Sight of the Big Picture

Whether in the crisis at hand, or with life in general. Check in with your colleagues. Check in with your partner, your family, your friends. *How am I doing?*

We all get wrapped up in our research. We all have demanding egos. And many of us are fiercely competitive. None of that makes you a bad person, and you shouldn't beat yourself up because of it. But you do have to be mindful of how these things impact you as a colleague, a family member, and an individual.

Don't lose sight of what's really important. If you are working on a crisis, are you making decisions based on what's best for your career, or what's best for the overall outcome? This is not an easy question to answer, because we are often blind to our own motivations and flaws. That's why it's crucial you check in with the people you work with, and with the people who love you. Listen to what they have to say. And then ask yourself again: what is truly important in this situation?

6. Process: Science Doesn't Move at the Same Speed as the Rest of the World

Even though science has enormous value to decision-makers during a crisis, the disconnect between scientists and those who respond to and are affected by such events is enormous. Time has become accelerated, and a ticking clock runs counter to every scientist's training. Contrary to popular perception, science is not a set of facts, but a method. Part of effective science communication is thus not only delivering content, but also reminding others how the scientific process works. Science was not built for speed.

At the same time, it is helpful to keep in mind what other stakeholders want from scientists—these expectations should help form your message. If you don't give people the answers they are looking for, someone else will. Responders care more about timeliness than accuracy. Reporters, always on a deadline, need stories that have a beginning, middle, and end. They want hard facts. The public wants to know: how does this affect me? If no answers are forthcoming, they will provide their own.

Perhaps one of the most useful sentences a scientist can say in a crisis is, "We don't have the answer for that yet." Lead with what you know is certain, but don't speculate and provide noninformation. If people want answers to something that is still an unknown, don't be afraid to hammer home the message: *We don't know yet, but the right people and resources are there ... it just takes time.* Stating that science hasn't found a puzzle piece, but is working on it, is just as important as communicating what you do know.

Journalism, the first draft of history, is incremental, too. Consider each scientific report like a chapter in an epic novel, and not necessarily in order. Let the dust settle and read the book in a few years.[8]

7. Misinformation: Remember to Listen and Display Empathy when Confronted with Alternate Viewpoints

Sometimes, it's going to require all of your available willpower to not explode with frustration. "Just look at the data!" you cry. But this is not how you are going to persuade someone who has already made up their mind. Engaging with a science denier is not about proving that you are right, or that you are more qualified than them. It's about getting the other person to open up to you. What is driving their belief in a conspiracy theory? What is really troubling them? What are they really afraid of? In order to reach someone, first you need to listen respectfully to what they have to say. You need to establish a person-to-person connection, as equals. If the conversation isn't going anywhere, then disengage before you lose your cool and make matters worse. Falling back on your credentials and demeaning the person you're speaking with isn't helping anyone; it's just cementing a stereotype.

At the same time, if you do have the data to back up your arguments, you should share it—somewhere, somehow—in a nonconfrontational way. A social media post or an op-ed may not get the facts to people who don't want to hear them. But it might get the facts to someone within their community—someone whom they are more likely to trust. And don't be afraid to follow-up on a conversation. People will appreciate your effort, so long as you don't come across as a jerk.

8. Legal: Don't Fight Battles You Can't Win—But Don't Stop Doing the Science

Regrettably, legal harassment and intimidation are becoming all too common in the scientific world. When my colleague Rich Camilli and I were dragged into the US government's lawsuit against BP, we tried to stick up for science. We wanted to protect our right to the deliberative process.

But ultimately, the cost of standing up to a multinational corporation like BP was just too high. Not only did we not win the battle, it ruined my life for several years. That said, if I had to do it all over again, I would have made the same decisions. I would have volunteered my time and efforts on behalf of the government. I would have published a peer-reviewed paper on our findings. The only thing I would have changed is how I dealt with that subpoena. I would have swallowed my pride and simply handed over everything that BP asked for. Even though it didn't feel like the right thing to do, the personal cost of trying to resist was just too high.

9. Impact: Before Any Action, Ask Yourself, "What Is a Successful Outcome?"

Every interaction is an opportunity. An opportunity to make an impact. The question is, do you want that impact to be a positive or a negative? There are many things to consider here, but perhaps none is more important than the idea of nourishing the person you are speaking with. Knowing what *you* want is easy— the key is figuring out what they want, and how you can both achieve your goals.

If you only focus on yourself, you risk leaving the other person unsatisfied. When I was taping an interview for *Good Morning America*, I fought with the journalist the entire time. The end result? I embarrassed myself and we both walked away unhappy. This was a missed opportunity. I should have prepared for that interview. If I had done my homework, I would have anticipated his angle and come up with a message that got my point across, without leaving him unfulfilled. But I was run down, and sometimes mistakes do happen.

That leads to another point worth remembering: nobody likes surprises. If you have bad news to share, make sure you tell the person who is on the receiving end ahead of time, so that they don't hear it from someone else or at the last second.

10. Teamwork: You Don't Need a PhD to Be an Expert

We all know that good communication is critical when working as a team. But the reverse is also true: teamwork is critical to good communication. Thankfully, scientists are already accustomed to working in teams. So much of our research, and so many of our papers, are collaborative efforts.

However, teamwork goes beyond just working with like-minded peers. We also have to learn to work with other professionals: graphic artists, writers, photographers, your PR department, marketers, web designers, professional responders and industry consultants who don't have their own lab ... everyone has a role to play.

A successful outcome means that you have communicated your content in a way that the intended audience not only understands the message, but also remembers it. Perhaps your audience goes on to share a graphic or fact sheet that was part of your message with *their* audience. That's when you know you've hit the jackpot.

As a scientist, you are responsible for the content. But everything else—the packaging, the platform, the design? Seek advice or hire a professional. And trust them to do their job. Even though they are not trained scientists, it doesn't mean that they won't understand what you are trying to say.

Just as importantly, when you find a person who does a good job, make sure to nourish that relationship. Tell them that you were happy with their work. Keep in touch, so that the next time you need professional help, you can go back to them. Because learning to trust other professionals can be the difference between a forgettable communications strategy and a successful outcome.

There is one final point I'd like to talk about. As we all know, the path to becoming a scientist is long and arduous. Doing all the coursework, passing your qualifying exams, and defending your dissertation are just the beginning. In the twenty-first century, having a successful career in the sciences is no sure thing—you need to be able to secure funding, be innovative, publish papers, spend thankless hours in the lab, and handle all the competitive pressures to boot.

With all that, it can be easy to lose yourself in your research. To shut out the rest of the world so that you can focus on what matters most. So on that note, I'd like to offer one last piece of advice: *don't let science define who you are.*

Colin Powell once wrote, "Avoid having your ego so close to your position that when your position falls, your ego goes with it." Yes, science is important, as is having a successful career. But you are more than just a scientist. You are a complex, multifaceted person, with other responsibilities, interests, and loves. Don't forget to nourish those other parts of yourself too.

I once knew a guy who was so obsessed with science that every night at the dinner table, he would lecture his kids on the subject. After a while, they stopped listening to him. By the time they had reached their teens they hated him. He never got those years back, and his original intention—to share his enthusiasm for science with his children—was a complete and utter failure. Unlike an interview, where you can always take another shot, life has no do-overs.

Finding a balance is tricky for everyone, and as we get older and accumulate more and more responsibilities, the harder it becomes. It might be tempting to hide from the emotional rollercoaster of day-to-day life in the lab. But it's only in living our life that we find our greatest sources of joy and inspiration. That we learn to connect the dots between what we know and what matters to the people we care about. That we can make our science matter. That we can finally answer the question: *so what?*

Notes

1 Minor changes to his calculation are the result of refinements to the variables in his calculus.

2 The first tetraethyl lead plant was known as the Butterfly Factory due to the number of workers who suffered from hallucinations of being attacked by insects.

3 It should be noted that the transition to unleaded gasoline was also spurred on by the introduction of catalytic converters that same year. Catalytic converters, which reduce car exhaust pollutants such as nitrogen oxides (NO_x), were rendered ineffective when used in conjunction with leaded gasoline because the lead in the exhaust coated the inside of the converter.

4 The research of Herbert Needleman, which demonstrated a direct correlation between lead poisoning and lower IQs and hyperactivity in children, was also instrumental in informing the public of the adverse health effects of lead. Needleman also became aware of the problem of widespread lead poisoning in the 1950s through his work as a pediatrician, and he campaigned for its removal from homes in the decades that followed. His landmark results on the effects of lead poisoning on children were published in the *New England Journal of Medicine* in 1979.

5 Timothy Dignam et al., "Control of Lead Sources in the United States, 1970–2017: Public Health Progress and Current Challenges to Eliminating Lead Exposure," *Journal of Public Health Management and Practice* 25 (January–February 2019): S13–S22, https://doi.org/10.1097/PHH.0000000000000889. Emphasis mine. It is important to note that not all communities in the United States are free of lead. Washington, DC, and Flint, Michigan, among others, are well-known examples of municipalities where lead pipes are still in use. Although efforts have been undertaken to replace municipal water lines, "the cost of replacement of private lines connecting homes to the public water supply and distribution most often is borne by the homeowner, which contributes to potential inequitable distribution of exposure. Nationally, an estimated 6.1 million lead-containing service lines exist today" (Dignam et al. 2019).

6 Megan Brennan, "40% of Americans Believe in Creationism," *Gallup*, July 26, 2019, https://news.gallup.com/poll/261680/americans-believe-creationism.aspx.

7 A staffer on the Hill once told me, "Prestige and status matter. Even if you are the best in the world but look too inexperienced or too young, we will not ask you to provide an in-person testimony."

8 C. Reddy, "How Reporters Mangle Science on Gulf Oil," *CNN*, August 25, 2010.

Appendix

How Science Works

Throughout my career, I have found that the challenges of communicating science are often rooted in not understanding how science works. Additionally, many of the people I have worked with outside the scientific community have a negative and inaccurate perception of scientists as a whole. Countless times, people have asked me questions like the following:

"Why can't you give me a straight answer?"
"Can't you just write a proposal to get some funding?"
"What is it with peer review?"
"Why are you so cautious?"
"Why are you so slow?"

When I say that people don't understand how science works, I'm not talking about the scientific method. Rather, I am speaking of the unspoken rules and values that govern our day-to-day lives and connect us as a community. As academic scientists, we tend to communicate based on the assumption that other people know where we're coming from. But it's all too easy to forget that most people have no idea what our lives are like. The complexities of funding, peer review, and tenure are just some of the stumbling blocks that can lead to inaccuracies and frustration.

In this Appendix, I highlight the key issues that are at the heart of many misunderstandings. For the scientist, this chapter serves to remind us that the inner workings of science are confusing and not necessarily self-evident. For nonscientists, this is the place to start if you want—or need—to understand our motivations. Why do we act the way we do? Why do we speak the way we speak? Why are we sometimes so hesitant to engage and share information? Why, when we do share information, do we tend to complicate things with too many details when people just want to hear more about the big picture?

The Baseball Analogy

Whenever I think about how to explain science to someone else, my mind immediately turns to baseball. What if an alien landed on earth and, for whatever reason, decided that it wanted to understand how this particular game worked? Would you sit the alien down on your couch, pull out the official rulebook, and read it out loud from start to finish?

Of course not. The alien would quickly lose interest and probably go on to eat you before you got to page five. Better to simply say: The object is to score points. You score points by running around the bases. If you get caught three times, it's the other team's turn. You might also want to mention that the game is slow.

There is no need to talk about infield flies, triple plays, stand-up doubles, suicide squeezes, or balks. The alien is not interested in the different types of wood used to make bats or how fast the ball moves after being hit. And yet, too often, I see scientists talk about just these things instead of saying, "You score points by running around the bases." We love minute details, however arcane they may be, and we love explaining them in great depth, because we assume that everyone else will find them just as fascinating as we do.

So here's a gentle reminder: if you find yourself talking at length about the infield fly rule, you're probably going to lose your audience. They will either tune you out or (worst-case scenario) eat you for dinner.

When you're partnering with outside stakeholders during a crisis—when the people who normally don't care about academia are suddenly forced to care—you need to be able to explain the rules that govern your world in a quick, clear, and concise manner. This can be particularly challenging, because in the heat of the moment, the values that govern your decision-making may be so deeply ingrained that you have a hard time articulating them.

If I had to focus on the key themes that govern our world, I would choose:

- The career ladder
- Peer approval
- Peer review
- Funding
- Publishing
- Certainty
- Running a lab

The order of this list and its exact content will be a bit different for everyone, so don't get too hung up on what goes where. These are just general guidelines.

What's important is to keep in mind the following: *why* do you do what you do? For instance, why are you a professor when you could likely work shorter hours for higher pay in industry? This question underlies everything else. Think back to Chapter 9: in the same way that trying to figure out what makes other people excited is a good conversation skill, remembering what makes *you* excited is equally important. Why do you wake up in the morning? What are you trying to solve?

While academics are a self-selecting bunch, we have a lot in common with many others in the workforce. We wish we were paid more and we work too much. At the same time, we enjoy following our curiosity and fulfilling our sense of discovery. We are relatively free to follow our research interests. People on the outside don't always understand this, but reminding them of the similarities and differences is a good place to start.

Before I dive deeper into each of these points, it is important to appreciate that career paths differ greatly. Academia is far from an even playing field— not everyone gets the same opportunities or the same access to training. As one female colleague explained, "I can say personally that my own path has been very different than my male colleagues'—for example, getting 'peer approval,' the amount of service that's expected of me, and having to deal with derogatory or mean reviews and comments."

The Career Ladder

So we have a passion—that's the why. The next logical question might be *how*. How does someone turn a potentially obscure topic (the sex life of the western tanager, the chemical composition of subsurface hydrocarbon plumes) into a career? For a scientist, this is obvious—this is what we've spent our whole life doing. But for everybody else, our career trajectory may be a bit confusing.

In North America, it's fair to say that there is one single pivot point that dominates an academic's career: tenure. Tenure is so important because it casts such a long shadow, and this shadow touches on so many parts of our lives. And because there is no equivalent of tenure outside the academic world—no one else in modern society has a guaranteed job for life[1]—it's hard for outsiders to grasp exactly how much influence it exerts, and how much social status is attached to it. Thus, when trying to explain a scientist's career path, tenure can be a good talking point.

When a journalist writes a book or provides training on science communication, they might make an argument along the lines of, "You, the scientist, have a moral responsibility to educate and inform the public. This is how you can use the media to do so in an effective manner." But this line of thought is ignoring a very important barrier: office politics. Do we want the public to be more scientifically literate? Absolutely. But do research universities and senior scientists want junior scientists to spend time talking with the media? Not so much. And this is important, because these are precisely the people who have total power over a scientist's future.

The first step in getting tenure—getting a job as an assistant professor— may be the hardest step of all: a successful candidate might need to beat out over two hundred other well-qualified applicants. If you've just completed your doctorate, that won't be enough: you need to have at least one year as a postdoc, and preferably you are already a visiting professor somewhere else.

You also need to have an impressive research portfolio. A recent study showed that a just-hired assistant professor of psychology had to have on average eleven publications, including at least three as the primary author.[2]

In today's academic landscape, it's much more common for recent graduates to serve as a visiting or adjunct (part-time) professor: at four-year universities, an estimated two-thirds of faculty positions are now non-tenure track.[3] Non-tenure-track positions typically receive low pay, few to no benefits, and have significant teaching loads that are prohibitive to doing research. According to a recent report from the American Federation of Teachers, nearly 33% of adjunct professors make less than $25,000 per year, while another 33% make less than $50,000.[4] What does this mean? That a junior scientist knows all too well that their grip on the lower rungs is extremely tenuous. While they may be personally motivated to engage with the general public, breaking ranks with academic tradition can be an extremely risky career move.

But let's say you are one of the lucky few who manages to become an assistant professor. The next few years are crucial to making the leap to tenured professor. Every university is different, but on average, a tenure review comes up around year six. During this time, you are expected to have proven yourself in the three main pillars of academic life: research, teaching, and service. Each university breaks down its expectations in different ways, but the 40–40–20 rule is often cited as an ideal. That is, a professor is expected to devote equal amounts of time to both research and teaching (two days a week), with an additional day devoted to service.[5]

The reality for scientists at major research universities (R1 or R2), however, is that teaching and service are not considered to be as important—and yet, they can still take up a tremendous amount of time and energy. As the deputy provost of Cornell wrote in 2022:

> Avoid over-investing in aspects of the job simply because they are the most personally satisfying or offer the most immediate gratification or positive feedback, or conversely, from devoting insufficient time to them because they are difficult or frustrating. These potential imbalances in effort can take many forms, but one of the most common happens regarding teaching and advising. Some new faculty members enjoy quick and satisfying success on this front. The energy of the classroom and the adulation of students can be gratifying, but your success here will not offset shortcomings in research and scholarship.[6]

For an academic scientist, what's important is the research you do, in addition to how much funding you obtain and how many PhD students you bring in. A typical allocation of work for an R1 university might be more along the lines of 60–30–10. With regard to service, many people, like Cornell's deputy provost, will tell you to do as little as possible. My advice is slightly different: go ahead and do service, but have an honest understanding of its impact within the system. If you are a woman or a minority, you will likely be asked to serve on multiple committees in order to provide diversity, so it is vitally important that you do not say "yes" to every request.[7]

Remember: scientists need to focus on their research. Not only does it need to be excellent, but they also need their colleagues to recognize its value. Because, ultimately, a scientist's peers are the people who are evaluating the tenure package. A tenure review will consist of both an internal evaluation (the provost, dean, and department chair and faculty) and an external evaluation (other specialists in the field). All of these people will review the candidate's portfolio.

Box A.1 Reddy's Rules

These are the ten rules that help me to stay on message when I work with outside stakeholder groups.

1 *Lead with certainty*: People are more interested in what you know, not what you don't know. Avoid making speculations in public.
2 *Always ask yourself, "What is a successful outcome?"*: Make sure you have a good understanding of your risks and rewards before you engage with other stakeholder groups.
3 *Mutually beneficial relationships are more than a first date*: Follow up with the people you meet. Relationships with people outside of academia are not one-offs.
4 *Look at an issue from all sides*: Learn to appreciate the perspectives of other stakeholders, even when you're unhappy with the outcome.
5 *Be a team player*: Scientists are sometimes brought on as "sages" or the "brain trust," with the expectation that our opinions carry more weight. But remember: we are equals and need to work with others in order to get the job done.
6 *Engage, don't lecture*: You've heard it before and I'll say it again. No one likes monologues. Have a conversation.
7 *Don't leave your audience hungry*: If somebody asks you to do something, do your best to fulfill their needs. For example, even if you can't answer a specific question, you might be able to point them to someone who can.
8 *Know your comfort zone*: Play to your strengths. If you aren't good on camera, don't go on camera. Don't be afraid to solicit feedback from others after a public appearance.
9 *Start big, go small*: When explaining how something works, don't read the rulebook from cover to cover. Instead, sum it up in a single paragraph.
10 *You can use jargon, but remember to explain what it means*: Jargon is not always the kiss of death, because people like to learn new things. However, when you use a specialized term or acronym, you must explain what you're talking about.

Peer Approval

Nonacademics might have a vague notion of what peer review is. But in academia, peer review is about much more than publishing. It's also about how your colleagues perceive you. So far, we've covered *why* and *how*; this is the *who* part of the equation. As we saw in the discussion of tenure, the epitome of a scientist's entire career is inextricably linked to their professional reputation. So what does this mean? Not only do you have to be a good scientist, you have to act the part. You have to follow the rules and support the value system.

Physicist Sean Carroll discussed this value system—and its relationship with getting tenure—in his cutting analysis, "How to Get Tenure at a Major Research University." This particular article sums up almost everything nonacademics need to know about the rules governing a scientist's life.[8]

For instance, it might be logical for outsiders to think that since universities are places of learning, the role of a professor is to serve as a public intellectual. However, as Carroll points out, this is not always the case. Major research universities can actually be staunchly anti-intellectual when it comes to anything not related to their employees' specialties. If you were hired to do X, you better not be dabbling in Y in your spare time, or, heaven forbid, writing a book (or even blogging) about Z.

In fact, writing a book is one of the cardinal sins he cautions against.

> If you're contemplating writing a popular book, and aren't sure whether it will negatively impact your chance of getting tenure, you're probably too far gone for this [article] to even help you. But it's worth a separate bullet point because even textbooks are beyond the pale. (Probably the worst thing I personally did was to write *Spacetime and Geometry.*) You might think that a long volume filled with equations that provides a real service to the community would help your case. It won't; it will hurt it. Why? Because while you were writing that book, *you weren't doing research.*[9]

In the same vein, he advises scientists to choose their hobbies wisely. (Cooking is fine, as is skydiving. But if you're thinking about dabbling in programming, inventing something useful, or starting a business? Keep it hidden.)

He also cautions against—stop me if you've heard this before—getting too much publicity.

> Don't be too well known outside the field. I hate to say this, but the evidence is there: if you have too high of a public profile, people look at you suspiciously. Actual quote: "I'm glad we didn't hire Dr. X; he spends too much time in the *New York Times* and not enough time in the lab." And that's the point—it's not that people are jealous that you are popular, it's that they are suspicious you care about publicity more than you do about research.[10]

Now, I am clearly not following all of his advice—I've written op-eds, been interviewed in the media on multiple occasions, and have even (eek!) written a book—but, I have been lucky; it's been twelve years since my last and final promotion. I also have several colleagues/collaborators who are carrying some of my weight. Simply, it was good fortune and good timing.

While the landscape is slowly changing, in part thanks to the social and technological developments of the twenty-first century, the fact remains: the scientific community can be a judgmental one. We live in an unforgiving environment, and our institutional memory is long. If you aren't going to follow the rules or meet traditional expectations, then you should have a full understanding of what the potential consequences will be. For some people, changing the system incrementally so that it is fairer and more inclusive is more important than getting tenure.

But even though increasing numbers of people are critical of the tenure process, we shouldn't apologize or feel less about ourselves because we're sticking to the rules of our game. Tenure exists for a reason: as Dr. Molly Worthen wrote, it provides "protection from arbitrary firing and retribution; it safeguards academic freedom; it decreases turnover and creates a more stable learning environment for students; [and] it's more cost-effective than critics suggest."[11] The bottom line is this: as scientists, we have our own value system and code of honor.

And in certain situations, we need to remind nonacademics of this. You might simply need to say, "Hey, this situation makes me uncomfortable, and here's why. I don't want to accidentally jeopardize my career or appear like a fool in front of my colleagues." Everyone has their own vocation-specific guidelines that they have to follow. Chances are, if you tell someone why you are reluctant to do something a certain way, they'll understand. Nevertheless, you still need to be careful—particularly during a crisis, when the stakes are high. What you say can always be taken out of context and come back to bite you, even years later.

Peer Review

A few years ago, I was having a discussion with my wife, Bryce, who wanted me to read a book on childcare. However, I wasn't sure what made this particular person the authority on the subject, so I asked her, "Who is this guy? Has his book been peer reviewed?"

My wife thought I was being a snob—that I wasn't going to take him seriously until I knew where he went to school, what he majored in, who was his PhD advisor, and so on. "But," I countered, "if the book hasn't been peer reviewed, how do you know what he says is reasonable and well supported? He could just be giving advice without having any data to back up his claims!" And then, for good measure, I added: "Anybody can write a book!"

Needless to say, this argument did not go over well, because Bryce had already read the book and decided it had value. In this case, *she* was the peer reviewer, and as a husband, I was discrediting her by casting doubt on her recommendation.

But the fact remains: as a scientist, I don't think I was off-base in wondering if his book had been vetted by other experts. I was just being rigorous, a quality that is instilled in every scientist over the course of their career. After all, childcare is an important subject that should be backed up with verifiable studies, not just personal conjecture. For people outside of academia, expertise might be conveyed by appearing on TV or being interviewed in the mainstream media. But for a scientist, the gold standard is a peer-reviewed paper. The fact is, being quoted in the *New York Times* carries no weight when it comes to credibility. When we want to determine if someone is legit, we check their publication record in academic journals. We look at their *h*-index (like it or not) or publication record on Google Scholar.[12]

For me, this anecdote is a perfect illustration of the gap between scientists and nonscientists with regard to peer review. My wife thought that I was being an elitist and casting doubt on her judgment. But for a scientist, the concept of having other experts review your ideas is enormously important. It is so ingrained in our DNA that we cannot imagine a situation in which peer review does not take place. We always expect someone else to challenge our assumptions. If they didn't, how would we know whether our ideas hold water or not? Questions, evaluations, and debate within the community are how science moves forward, and how inaccuracies are weeded out.[13]

But we're getting ahead of ourselves. The first thing a nonscientist may be wondering is: how does the peer-review process work? So let's start from the beginning.

Publishing a Paper: It All Starts with Funding

Let's say a scientist identifies a topic they're interested in researching. The first thing they have to do is write a detailed proposal to obtain the funding necessary to carry out their research. Like everyone else, money plays a big role in what a scientist wants to do and what they get to do. Writing the proposal takes at least a month. Next, they submit the proposal to a funding organization like the National Science Foundation (NSF)—it's worth noting that the application window for some organizations is open twice a year, while for others it may only open once a year.

After the funding organization receives the proposal and checks it for compliance, they forward it to three external experts in the field, who review it anonymously. This is called a letter review. After the letter review, the proposal is sent to the funding organization's panel review. The panel consists of ten to fifteen scientists who review all the proposals that have been submitted within the relevant field for that submission window. So, to recap: before your project can even begin, over a dozen people critically evaluate and then grade it. And here's the kicker: most proposals are not funded. While outsiders sometimes think of funding as a painless procedure where scientists ask for money and then receive a blank check, it is in fact a long, laborious process with a low success rate.

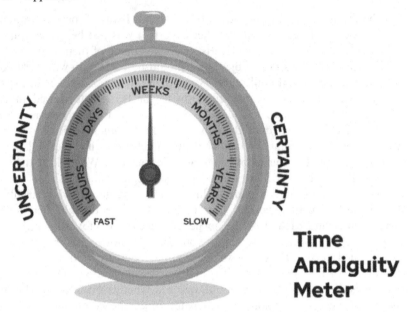

Figure A.1 In the earliest stages of a crisis, information may carry significant uncertainty. Responders are willing to accept uncertainty to do their jobs. In academia, careers can go bust with too much uncertainty.

In my field, roughly 25% of proposals receive funding; the remainder can try again, making revisions based on the reviewers' comments. When this happens, the scientists can't resubmit a revised proposal until the next submission window opens up, which could be months away. In a worst-case scenario, you may need to come up with an entirely new line of study. Since most people don't get their proposal funded the first time around, this means you have to repeat the process until you succeed, give up, or get a better opportunity. Even in a best-case scenario, going from the kernel of an idea to receiving funds in your institution's bank account will take at least six months, and often takes much longer.

As one of my colleagues stated:

> I think most of us have dozens of proposals whose research never came to be, for whatever reason—high risk, bad luck with multiple submissions, or even proposing a line of research that's too interdisciplinary so you end up with fifteen reviewers for the proposal! It's important to understand that even successful people struggle and that in the United States, the funding model is one of the biggest bottlenecks to actually advancing science—or at least advancing it faster.

From Research to Publication

Once a project is funded, the next step is to carry out the research. We can kid ourselves that we will be in the lab all day, but in reality, we have to leave it to others. The first thing you do is staff your lab with a mix of graduate students, postdocs, and lab technicians. Then you start running your experiments; in order to write a paper, you might need anywhere from five months to five years to accumulate enough data. Finally, you're ready to write your paper. A paper is usually anywhere from three thousand to six thousand words in length, and will contain a couple of figures and tables. Altogether, writing the paper will take you at least three to four months.

Next, you write your cover letter and submit the paper to a journal. The editor might look at it and say, "Not for me," and send it right back to you. Then you have to find someone else. Or they may say, "This looks kind of interesting." In the latter case, they send it out to three other scientists who will evaluate it. The other scientists might take anywhere from one to three months to send back their comments.

The editor reads everything and comes to a decision. Most decisions are either along the lines of, "You're accepted, but you need to make some changes" or, "Sorry, not interested." If you're accepted, it might take you a few more months to make the requested changes and rewrite the paper. The paper then goes back to the editor.

They might send it out to one other scientist just to double-check. After that, your paper is finally ready to go. So, best case scenario, you are looking at two years of work—and this is after you spent a year trying to get funding. Can you publish a paper in less time? Sure, but it is unusual. This is the nature of science: things move slowly.

Certainty

One of the most frustrating things when writing a paper is that you think you have a solid story. You think that your conclusions are at about 90% certainty. But then you find out there's a flaw and you're only at 65% certainty. That either means a do-over or a lot of work, because you know that peer reviewers do not feel comfortable with papers that don't have a high level of certainty (and neither do you!).

Uncertainty is one of the things that makes scientists particularly nervous. We always think we have to minimize the uncertainty. We have to be absolutely confident about what we do, because if we aren't doing our absolute best, then we're not going to meet the standards that our peers set for us.

And this is another point of friction when scientists are working with other stakeholder groups. On multiple occasions, the response community has told me, "I don't care about peer review. I don't care about certainty. What I want is for you to tell me something that I can use in my decision tree. Even if you think that you're only 75% certain, bring it on."

Crisis responders sometimes find us too rigid, because we are overly concerned with our own output and whether or not our colleagues will find our work acceptable. Some have not hesitated to critique us, saying things along the lines of:

> All you care about is the paper you're writing and how you present the science. You think that science is only valuable when it is published in a peer-reviewed journal. But you can add science to the table that isn't peer reviewed and we'll value it. I'd be psyched if I could get your professional insight on a back-of-the-envelope calculation.

Time and certainty go together like hand and glove. We are used to working in an environment that moves slowly, and operates with a high degree of certainty. But sometimes, during an environmental crisis, the rules of the game are accelerated, and we need to accept that there may be more than one way to communicate essential data (e.g., the FAQs and fact sheets discussed in Chapter 10).

Running a Lab

There's one last point I want to touch on: running a lab. When explaining what scientists do in their day-to-day life, it's important to remind others that professors at major R1 universities are not there to teach. We might be assigned one class and expected to advise a certain number of students, but our value to our employer is as a researcher. To put it another way: the university provides us with a place to work and pays us a nine-month salary. Our part of the bargain is not to teach undergrads, but to bring in outside funding. As one of my colleagues wrote on Twitter, there are three things that every scientist wants: funds, PhD students, and papers. "The first one creates the domino effect for the last two."[14] If a scientist is unable to secure funding for their research program, they will wind up with an empty lab, no PhD students, and no papers.

That's why I always tell people the following: the best way to think of a scientist is as a small-business owner. The universities where we work are giant malls, and each researcher is responsible for running their own store inside that mall. The Nobel Prize winner has the anchor store, and the rest of us are scattered around in the smaller shops. The US government is the investor, allowing us to produce papers (our product) for our colleagues (our customers). Just like a start-up, we have to constantly hustle to drum up investment dollars.[15]

This is important, because while it's the product that ultimately determines our value, we need to be able to keep the lights on for long enough to create that product. And this, as we know, often takes years. For scientists, this also presents a Catch-22 scenario: sometimes, in order to get funding, you need to provide preliminary data. But in order to get the preliminary data, you need the funding. How does one get started?

The complex problem of securing research money came under new scrutiny during the early days of COVID-19. As Derek Thompson recounted in the *Atlantic*:

> In April 2020, when the coronavirus first swept across the United States, many of America's top scientists struggled to get funding to answer basic and urgent questions about the disease it caused. Patrick Collison, the chief executive of the payment-processing company Stripe, spied an opportunity in this market failure. He co-founded a program called Fast Grants, which raised more than $50 million that was quickly distributed to hundreds of projects. ... The success of Fast Grants raised an uncomfortable question about how the U.S. funds innovation. If a little pop-up could unlock so many good ideas so quickly, how many potential breakthroughs are being denied every year by the traditional system of funding science? ... Bureaucracies move slowly and require arduous busywork. Today, researchers spend 10 to 40% of their time putting together complex grant proposals. This time suck pulls scientists away from doing real science while it nudges them toward projects that will appeal to peer-review boards rather than lead to novel breakthroughs.[16]

We all know that Silicon Valley-style disruption does not always yield the best results: Theranos is the most egregious example of a STEM start-up gone wrong. But this article identified one of science's big problems going forward: why is it so hard to obtain funding, especially for new, untested ideas?

We shouldn't have to rely on venture capitalists to unlock new ideas and fund research. In my own experience during the *Deepwater Horizon* crisis, scientists were able to obtain RAPID grants—which offer much less money, and which also have no peer review component—to get out in the field and do research within days of submitting a proposal. It was chaotic. Many people were left with a bitter taste in their mouths. But, at the same time, the science produced from those grants was incredibly validating.

When communicating with new people, we want to look for overlapping interests and shared experiences. Explaining to people that as a research scientist you are operating a small business is a great, reliable way to make a connection. Because, just like business owners, scientists also worry about their revenue stream, buyer interest, production time and quality, reputation, and sustainability.

Notes

1 Federal judges are a notable exception.
2 D. Reinero, "The Path to Professorship by the Numbers and Why Mentorship Matters," *Behavioural and Social Sciences*, October 23, 2019, https://socialsciences. nature.com/posts/55118-the-path-to-professorship-by-the-numbers-and-why-mentor ship-matters.

3 This calculation includes graduate students who serve as TAs. AAUP, "Data Snapshot: Contingent Faculty in U.S. Higher Ed," October 11, 2018, www.aaup.org/news/data-snapshot-contingent-faculty-us-higher-ed.

4 C. Flaherty, "Barely Getting By," *Inside Higher Ed*, April 20, 2020, www.insidehighered.com/news/2020/04/20/new-report-says-many-adjuncts-make-less-3500-course-and-25000-year.

5 See H. Dixon and H. Tervanotko, "How Do Tenure-Track Professors Really Spend Their Work Time?," *Chronicle of Higher Education*, December 14, 2021, www.chronicle.com/article/how-do-tenure-track-professors-really-spend-their-work-time.

6 John Siliciano, "A Short Guide to the Tenure Process," Cornell Engineering Department, www.engineering.cornell.edu/research-and-faculty/faculty/resources-faculty/faculty-development/tenure-track-faculty/promotion.

7 K. Kelsky, "The Professor Is In: Service Taking Up Too Much Time?," *Chronicle of Higher Education*, June 26, 2017, www.chronicle.com/article/the-professor-is-in-service-taking-up-too-much-time.

8 S. Carroll, "How to Get Tenure at a Major Research University," *Discover*, March 30, 2011, www.discovermagazine.com/the-sciences/how-to-get-tenure-at-a-major-research-university.

9 Carroll, "How to Get Tenure." Emphasis is mine. Carroll did not get tenure at the University of Chicago, which he attributes in part to the fact that he wrote the aforementioned textbook while serving as a professor. He went on to take a position as research professor at Cal Tech, and is currently the Homewood Professor of Natural Philosophy at Johns Hopkins.

10 Carroll, "How to Get Tenure."

11 M. Worthen, "The Fight over Tenure Is Not Really about Tenure," *New York Times*, September 20, 2021.

12 The *h*-index is not a perfect system and it has its share of critics. One of the biggest problems, of course, is that it favors senior scientists—they have had more time to publish and hence be cited. Some people argue that the community should use the formula "h score/x," where *x* is the number of years since your first publication. This levels the playing field somewhat. For more critiques of the *h*-index, see G. Conroy, "What's Wrong with the h-Index, According to Its Inventor," *Nature*, March 24, 2020, www.nature.com/nature-index/news-blog/whats-wrong-with-the-h-index-according-to-its-inventor.

Nevertheless, the *h*-index does have value and I will sometimes point the media or a policymaker to Google Scholar, in order to argue that a certain scientist is either a real star or not the expert they are portrayed as.

13 The flip side to this is that some scientists occasionally make it personal and attack others, either because they feel threatened by new research or because they're holding a personal grudge. When the peer reviewer is a senior scientist and the person being challenged is a woman, a person of color, or a junior scientist, power dynamics can become a serious issue.

14 @numericalguy, Twitter post, August 24, 2022, 8.58 p.m., https://twitter.com/numericalguy/status/1562635410768945152.

15 What sets us apart from the small business model is that our business is also helping other businesses find success, helping our employees find new and better jobs, and reviewing other businesses' products. However, like a small business, we still have to cover overhead. In my case, roughly one-third of my research funds are used to pay for all the associated costs of running a lab.

16 D. Thompson, "Silicon Valley's New Obsession," *Atlantic*, January 20, 2022, www.theatlantic.com/ideas/archive/2022/01/scientific-funding-is-broken-can-silicon-valley-fix-it/621295.

Index

Printed in the United States
by Baker & Taylor Publisher Services